DARWINISM

and the

LINGUISTIC

IMAGE

New Studies in American Intellectual and Cultural History

Dorothy Ross and Ken Cmiel, Series Editors

DARWINISM
and the
LINGUISTIC
IMAGE

Language, Race, and Natural Theology
in the Nineteenth Century

Stephen G. Alter

The Johns Hopkins University Press
Baltimore and London

© 1999 The Johns Hopkins University Press
All rights reserved. Published 1999
Printed in the United States of America on acid-free paper
2 4 6 8 9 7 5 3 1

The Johns Hopkins University Press
2715 North Charles Street
Baltimore, Maryland 21218-4363
www.press.jhu.edu

Library of Congress Cataloging-in-Publication Data will be found
at the end of this book.
A catalog record for this book is available from the British Library.

ISBN 0-8018-5882-8

For my parents,
the patient ones

CONTENTS

ILLUSTRATIONS

PREFACE AND
ACKNOWLEDGMENTS

This book traces a colorful metaphoric thread running through the early post-Darwinian debates in Britain and America. Charles Darwin and his peers had a surprising amount to say about a similarity they perceived between the transmutation of biological species and the "evolution" of languages. In calling attention to the patterns of historical development among certain words and the ties of kinship among certain groups of languages, some of the era's most prominent scientific writers wielded an impressive polemical tool, one with which they could both illuminate and interpret Darwin's theory of organic descent.

I first encountered this striking parallelism while researching a related topic: nineteenth-century Anglo-American linguistic theory. In that context, the comparison bespoke a strong interest in natural science, obviously including Darwinism, on the part of the founders of what then was called the science of language. These analogies signified a deeper resonance than did merely isolated cross-disciplinary illustrations, for they seemed to open up an entire world of integrative thinking among intellectuals of the Victorian era, an outlook that bid to unite the natural and human-cultural spheres. The question in this book, however, is not only how linguists looked at Darwinism but, more especially, how and why scientific thinkers took an interest in the quintessentially humanistic subject of the history of language.

Explaining the attraction that linguistic phenomena held for Darwin and showing how his fellow naturalists responded to the comparison would fill, I thought, a substantial article. Yet the instances of the analogy piled higher as I worked my way through the relevant writings—the books, articles, personal letters, and private journals, especially those from the height of the evolution controversy during the 1860s. It became clear that the scientists and language scholars of Darwin's generation were more heavily invested in this image than I had originally realized. So the story grew in the telling, and this book is the result.

The narrative proceeds in five stages. As a preliminary, I take stock of European linguistics in the pre-Darwinian period, a field that was itself already steeped in the imagery of the life and earth sciences. The story begins in earnest with the language-species analogies found in Darwin's early writings, then moves to the similar figures appearing in the works of his contemporaries in the decade after the appearance of *The Origin of Species* (1859). Next I explore a subtler form of polemic, in which some writers debated the Darwinian question by linguistic means alone. Whether overt or hidden, these appeals to the linguistic image had the ultimate effect of provoking a climactic response from Darwin in his *Descent of Man* (1871).

My final chapter and the epilogue, the most speculative part of my argument, are considerably wider in scope. Here I highlight the concept of genealogy as a central integrative idea in both the biological and human sciences in the late nineteenth and early twentieth century. At that time, a number of language-based disciplines adopted a comparative research method and produced a genealogical arrangement of their data, thereby mirroring the Darwinian tree of life. Another way to put this is to say that the briefly stated linguistic analogies used in the evolution debates foreshadowed a much larger convergence between the human and natural sciences. Ultimately, I suggest that this coming together of such disparate fields of study could not but have made an impression upon readers and would have reinforced the manifest plausibility of branching-descent patterns in whatever context they appeared. In this way, even apart from any conscious analogizing, linguistic study would have influenced popular scientific opinion, helping to create the climate of public receptivity that eventually was accorded Darwinism.

I add here a note on terminology: *Philology* was the most common English label for the linguistic field as a whole throughout the nineteenth century. Used more strictly, the term suggested the field's dominantly historical and often literary bent. *Comparative philology,* really a misnomer, described nearly the opposite: this was the English name for research into languages in isolation, apart from their literary and historical associations. Another term was *linguistic science* (based on the German *Sprachwissenschaft*), encompassing not only comparative research but general linguistic theory as well. The noun *linguistics,* after the French *linguistique,* also became popular in the latter part of the nineteenth century, sometimes used in contradistinction to traditional text-centered philology. Yet there was no agreement or consistency in the application of these names, and, despite the attempts to make distinctions, *philology* and *linguistics* were used nearly interchangeably until the early twentieth century. I continue this practice, for the sake of variety, except where I indicate that a special meaning is intended.

Many scholars, both friends and those I know only in writing, have generously helped in the production of this book. For providing me with information or for their critical reading of some portion of the manuscript, I thank Gillian Beer, Sandra Herbert, Henry Hoenigswald, Christopher Jones, Konrad Koerner, David Livingstone, Edward Manier, Stephen O. Murray, and Jon Roberts. Special thanks go to Donald Fleming, Thomas R. Trautmann, and James Turner. Each, at a crucial stage in this project, gave me the benefit of thoughtful criticism; more than this, each has exemplified on countless occasions the spirit of scholarly mentorship. It has been my privilege as well to be associated with the editors of this book series, model scholars in their own right. Dorothy Ross and Kenneth Cmiel pushed the big questions, made me pare down my prose, and helped me to clarify the argument. I could not have been better served by them or by Robert J. Brugger of the Johns Hopkins University Press, who lent to my efforts his wise editorial oversight. Of course, the responsibility for how I have used all of this advice rests with me, not with the various individuals consulted.

A dissertation grant from the Mellon Foundation, awarded in 1992–93, launched the initial phase of my research. Indispensable support also came from a year's fellowship in 1994–95, and continued hospitality thereafter, at the Charles Warren Center for Studies in American History at Harvard. I hope that this book can repay in some small measure the manifold practical helps and encouragement I have received from the Warren Center's directors, staff, and fellows. I also thank the staffs of the Gray Herbarium Library and Special Collections, the Museum of Comparative Zoology Library and Special Collections, and the Interlibrary Loan Service of Widener Library, all of Harvard University. Alexandra Oleson of the American Academy of Arts of Sciences and Rebecka Persson of the Boston Athenaeum cheerfully helped out by providing me with a copy of manuscript minutes of an 1861 American Academy meeting. Thomas Adams of Harvard Imaging Services rendered aid beyond the call of duty with the illustrations. A very special debt of thanks goes to Jon Topham of the Darwin Correspondence Project for his gracious and professional aid in securing unpublished letter transcripts from the Darwin collection at the Cambridge University Library. For permission to quote from their collections, I thank the Cambridge University Library, the Gray Herbarium Library of Harvard, the American Academy of Arts and Sciences, and the American Philosophical Society Library.

Finally, the thought expressed in the dedication of this book is one that writers and their loved ones will understand without benefit of explanation.

DARWINISM

and the

LINGUISTIC

IMAGE

Prologue

Science as Indirect Discourse

CHARLES DARWIN WAS NURTURED in a scientific milieu that cared much about the public presentation of its ideas. It was a care and an interest that extended even to the use of figurative language, including illustrative metaphor and analogy. Historians have paid considerable attention of late to such issues of scientific rhetoric, part of a larger effort to divest science of some of its privileged immunity and show that it, too, is a cultural activity.[1] What is especially interesting about Darwin's generation is their particular selection from the available nonscientific imagery.

Every age, perhaps, has its special predilections with regard to this kind of cross-disciplinary affinity, its couplings of different phenomena that mutually resonate nonetheless.[2] These seemingly natural metaphors—half-conscious bonds of logic among distinct fields of knowledge—draw upon the aesthetic sensibility of a given time and place: they ground the communicative strategies and plausibility structures of science in juxtapositions that are as much imaginative as they are cognitive.[3] Study of the illustrative figures used in past scientific discourse thus affords insight into larger habits of interdisciplinary transfer, a topic having as much intellectual-historical import as the study of scientific findings themselves.

The best-known sources of Darwin's own figurative thinking include Malthusian population theory, the division of labor as conceived by the classical economists, and selection as practiced by domestic plant and animal breeders. Leaving aside the last of these, one finds comparisons bridging distant conceptual boundaries, particularly implying a linkage between natural science and the organized study of society. Darwin's analogies with linguistic phenomena fit into this same general category, for those images represented nature through a quintessentially socio-cultural realm. By extension, the analogy included as well the human family's racial divisions, for language and ethnos had traditionally been paired with each other as covariants in human history.

Darwin and other scientific writers appealed to both classes of phenomena, the linguistic and the ethnological, in their attempts to illuminate the vexing question of species. These images functioned most basically in the way that figurative discourse always functions. In all use of analogy and metaphor, it is difference that allows one to really see; the other provides the ground of contrast that makes perception possible. Whether brief and simply stated or, as in some cases, wrought with impressive intricacy, parallelisms between nature and language served as imaginative attempts to influence the reception of Darwin's descent theory.[4]

The nineteenth century saw the heyday of philological scholarship and especially of the new comparative philology, the latter being a predecessor of the modern field of linguistics. Philology of all kinds was distinguished by its historical emphasis, its concern with linguistic and literary change over time. Although a connection with science here may seem unlikely, the conceptual link was already in place. Certainly in Britain, at least, the philologist and the antiquary were related from the start, and the latter was brother to the (often amateur) geologist and paleontologist.[5] The retrospective aspect of all these activities placed on a par the collectors and interpreters of old manuscripts, Roman relics, pre-Cambrian rocks, and, eventually, prehistoric fossils; the interpretation of all such artifacts demanded that they be arranged in a temporal series. It was this historical quality, most basically, that allowed philology to lend itself so readily to metaphors of organic growth and ultimately to comparisons with biological evolution: the slow transformation of languages provided an apt analogy for the gradual transmutation of species.

Yet what was especially telling in relation to Darwinism was the added comparativist dimension: through the analysis of linguistic forms shared by an entire set of contemporaneous languages, scholars sought to determine the relative degrees of kinship among those tongues. Comparative philology thus embodied a distinctive kind of historical vision, one that was not merely genetic but genealogical.[6] That is, the new discipline was built around the idea of branching descent from a common ancestor.

Certainly no philologist himself, Charles Darwin devoted only a small number of passages in his writings, and most of these quite brief, to the discussion of language. His linguistic analogies, moreover, comprised but a fraction of the many nonbiological images he used. As the historian Susan Faye Cannon observed, Darwin "seized upon any metaphor, any analogies, any line of argument he had on hand" in order to convey his vision of the natural world's operations.[7] Even so, linguistic figures appear at several strategic points in *The Origin of Species* (1859), and an entire series of such comparisons embellishes an early chapter of *The Descent of Man* (1871). It is obvious that these images could offer no direct evidence of biological transmutation, which raises the question of precisely how they helped buttress Darwin's argument.

Before suggesting an answer, I should like to eliminate a potential source of confusion. One ought to distinguish between the language-species analogy and speculations about the actual origin of human speech. The latter issue attracted considerable interest during the nineteenth century and was pertinent in its own way to the Darwinian debates. The Oxford philologist F. Max Müller, for instance, said this soon after the appearance of *The Origin of Species*: "Language is our Rubicon, and no brute will dare to cross it. . . . It admits of no caviling, and no process of natural selection will ever distill significant words out of the notes of birds and the cries of beasts." Müller's pronouncement capped off a long tradition, begun decades prior to Darwin's famous book, of linguistically informed "natural theology," a polemic against the materialist view of biological existence. European writers routinely cited language as a principal barrier separating human from mere animal intelligence, and some used philological evidence to argue that speech must have had a transcendent origin.[8]

All of this, however, addressed the problem of the origin and nature of "man," while the linguistic analogy applied to species in general, from orchids to sperm whales. In other words, the analogy had nothing to do, necessarily, with the origin of human speech—although the two themes often became entwined in the heat of argument. Admittedly, there was a degree of conceptual overlap. Both the language-species analogy and the evolutionist theory of the origin of speech stressed the gradual emergence of complexity out of relative simplicity. Darwin himself, on several occasions, fudged his speculations about the emergence of articulate speech into descriptions of the ongoing development of language. Generally, however, his and others' emphasis on the similarities between linguistic change and biological transmutation did not address the question of the "evolution of man" or the initial emergence of speech but supported Darwin's case for evolution as a whole.[9]

Darwin clearly expected this analogy to help sway undecided or even hostile minds as they considered his descent theory. He intended it, that is, to enhance his theory's plausibility. More than mere colorful illustrations, linguistic images could link what was controversial and tentative to what was more familiar and convincing. In particular, they could habituate readers to thinking in terms of gradual transformation and branching descent; they could normalize the idea that a single prototypical ancestor might give rise, over the long term, to widely divergent offspring. As philologists were demonstrating at that time, patterns of "descent with modification" had indeed occurred among many of the world's tongues: the notion of an Indo-European family of languages offered only the best-known example. These kinds of discovery, it was thought, reduced the prima facie implausibility of something similar occurring among biological species. Darwin's analogies thus embodied an argumentative stratagem similar to that which the historian Martin Rudwick finds in Charles Lyell's writings. Lyell aimed to convert not only the scientific reason but also the "scientific imagination," particularly with reference to the vast time scale required for producing gradual geological change.[10] Rightly so, Darwin must have deemed such a conversion of the imagination even more essential to gaining the public's acceptance of biological evolution.

Charles Darwin was one of about ten major scholars of his day whose writings, public or private, invoked linguistic analogies. The geologist Lyell, Britain's chief spokesman for that burgeoning field, stood foremost within this group. Two of the others were, like Lyell, among Darwin's closest scientific friends—the American botanist Asa Gray and the British zoologist Thomas Henry Huxley. Louis Agassiz, a famous opponent of Darwinism, played a role as well, expressing his opposition not only through paleontological arguments but also in linguistic terms. On the philological side, contributors to the discussion included Darwin's cousin and brother-in-law, Hensleigh Wedgwood, the Anglican churchman Frederic William Farrar, and the Victorian world's most celebrated linguistic savant, the German-born Friedrich Max Müller. Finally, two university colleagues in Germany itself took part: the respected comparative linguist August Schleicher and the fiery young zoologist Ernst Haeckel.

Each of these figures possessed the kind of technical competence in his field which, although outmoded, can still be appreciated. More thoroughly alien to today's scholarly sensibility is the approach they took to scientific argumentation. The important fact is that these writers could expect to address, and were obliged to appeal to, a general readership. Of course, it was much less of an age of specialization, and so the makeup of the reading public, in addition to the classical education many scientists of their day still received, led Darwin and his peers to infuse their writing with a literary quality that professional scientific discourse has long since abandoned. By later standards, this imaginative element made relatively small things appear overly large—the case in point being the seriousness with which these writers took "mere" illustrations. Not only did they engage in sometimes extended discussions of natural history via purely analogic means, but they increasingly played off one another in this fashion, cleverly mirroring their private disagreements over the meaning and validity of Darwinism. In order for readers to have appreciated this indirect mode of discourse, they would have needed at least a rough acquaintance with contemporary linguistic research. As it happened, such familiarity was becoming widespread throughout the English-speaking world in the years surrounding the publication of Darwin's *Origin*.

Readers should be warned, however, that this knowledge is not in all respects accurate by today's lights. Mid-nineteenth-century philology held an overly schematic view of linguistic "descent" and was often mistaken in such details as individual word histories. Moreover, Darwin and his circle were philological laymen, writing for a general audience and using commonplace terms. Charles Lyell, for instance, described some languages as being new and declared that none of the modern European tongues was more than one thousand years old. Neither of these ideas was strictly correct, for languages never make completely new starts. In reality, old languages either evolve up to the present day or else die out. Still, one may accept remarks such as Lyell's in the spirit in which they were intended; one can regard modern Italian, for example, as a new language in comparison with Latin. Except when I suggest otherwise, I present the linguistic notions of the period, regardless of present scholarly merit, as writers then understood and expressed them.

Although linguistic themes could be made to fit several different facets of Darwin's theory, they reflected particularly well his idea of common descent. That is, they conveyed the notion of divergent speciation arising from less differentiated predecessor forms and ultimately from a single progenitor species. Writers also sometimes pointed out linguistic processes that seemed to parallel the idea of natural selection. In making this distinction, I follow the biologist Ernst Mayr, who finds a number of separate components within Darwin's overall theory: we may thus distinguish descent from natural selection conceptually.[11] It is equally important to distinguish these components historically, in terms of the actual career of Darwinian thought. As Peter Bowler shows in a number of his works, most naturalists came to accept biological descent in the post-*Origin* decades, even as they discounted natural selection as an adequate explanation of that process. Bowler notes as well a second crucial feature of the post-*Origin* period: that the idea of treelike branching added a new element to the old doctrine of biological transmutation. "Evolutionists" prior to Darwin, notably J.-B. Lamarck and Robert Chambers, had proposed unilinear rather than branching schemes of descent. Although he does not stress the point, Bowler thus suggests that there was something besides natural selection that was distinctive to the "Darwinian revolution" and that a crucial aspect of full-fledged Darwinism took hold soon after 1859.[12]

My point is that common descent, as such, constitutes a significant and understudied notion within the historical complex of ideas commonly called Darwinism.[13] Following Darwin's own practice, writers have usually represented this pattern of development through the figure of a branching tree; this in turn stood for a family tree, a picture of kinship relationship as well as of taxonomic classification. The notion of divergent descent is thus considerably more elaborate than the generic idea of biological evolution—even though, popularly, *evolution* has become nearly synonymous with the branching schema. This aspect of Darwinism, massive and obvious, ought to be examined more closely by students of the cultural history of science, its very familiarity questioned historically. The effort would involve searching through the intellectual environs of Darwin's lifetime and since, looking for the various resonances of the family tree metaphor. In short, we should do for the concept of common descent what has been done for the idea of natural selection.

The most sustained work in this area has been done by Howard E. Gruber, who examines the appeal that the branching tree figure held for Darwin. Building on Gruber's observations about the multiple meanings embedded in Darwin's tree metaphor, and emphasizing the similar pattern one finds in analogies of linguistic descent, I nevertheless differ on one point. Gruber argues that the tree image reflected the disorderliness of evolution, that it embodied for Darwin (probably unconsciously) an "aesthetic of complexity."[14] I suggest, on the contrary, that this image served a simplifying function in relation to both languages and species. This aesthetic of coherence, I argue, was not born of any absolute symmetry or regularity; Darwin's evolving nature was indeed the opposite. Rather, for Darwin nature's tree rooted wild proliferation in a basic unity; beneath the branching chaos, the common trunk secured a fundamental, indeed organic, integrity. This yearning for harmony, particularly the superimposition of the developmental patterns found in the biological world and in language, was symptomatic of the intellectual life of the age.

≈ 1 ≈

Comparative Philology and
Its Natural-Historical Imagery

THE METAPHORIC MIND of nineteenth-century science produced some striking conceptual transfers. One of these, the main topic of this book, is the way linguistic ideas helped convey that period's most advanced theory of biological nature. Yet this was really reciprocal, for well before the advent of Darwinism, linguistics looked to science for its own methodological model. Accordingly, this introductory chapter includes a brief survey of European language study in the century prior to the Darwinian revolution, focusing especially on that field's natural-historical self-presentation and imagery. Thus told, the story replicates comparative philology's self-congratulatory image as a "new science," not really invented until after 1800.

Of course, this picture obscures longer continuities, including the way comparative philology perpetuated aspects of a traditional, Bible-based view of the history of languages. It also side-steps nineteenth-century linguistic study's ethnological presuppositions, these again being rooted in the idea that peoples and their languages dispersed in tandem after the crisis at the Tower of Babel.[1] In short, this chapter emphasizes not the deep and complex roots from which early-nineteenth-century philology actually grew but the "scientific image" of the field fashioned in that period. This image took its cues, first of all, from German claims that the new wave in language study employed the methods of natural science. The British added to this perception by making linguistic research the analog particularly of geology and paleontology. These natural-historical themes were among the basic assumptions about language held by early Victorian thinkers. Hence they shaped the linguistic worldview of Charles Darwin and his peers.

The Indo-European Idea in Linguistics

The latter half of the eighteenth century saw a rising interest in the history and comparison of languages, which was born largely of a concern to map lines of ethnological kinship. One such approach to comparative linguistic study was that in which Western scholars compiled basic vocabularies of the world's tongues. These simple inventories eventually allowed languages to be grouped according to similarities of characteristic word structure. The champion of this structural mode of classification, and the figure who did the most in this era toward preparing an index of the world's tongues, was the German scholar-statesman Wilhelm von Humboldt (1767–1835). Humboldt's division of languages into four "morphological" types enjoyed a long scholarly life, although it became partly absorbed within, and partly overshadowed by, a newly emergent comparative-historical philology.

The impetus for this new field came with the realization that Europe's classical languages bore close genetic ties to Sanskrit, the ancient language of the Hindus. A number of researchers made this discovery independently, yet the most influential writer on the subject was the British jurist and colonial administrator Sir William Jones (1746–94). Jones's service in India required a close acquaintance with native legal traditions, and he became familiar with a number of "Oriental" tongues. In 1786, soon after helping found the Royal Asiatic Society of Bengal, he announced to that body his observations about linguistic common descent.

> The Sanskrit language, whatever may be its antiquity, is of a wonderful structure; more perfect than the Greek, more copious than the Latin, and more exquisitely refined than either; yet bearing to both of them a stronger affinity, both in the roots of verbs and in the forms of grammar, than could have been produced by accident; so strong that no philologer could examine the Sanskrit, Greek and Latin, without believing them to have sprung from some common source, which perhaps no longer exists.[2]

This finding laid the basis for the Darwinian analogy that came later, for it posited the common derivation of widely divergent phenomena and suggested that the ancestral progenitor had long been extinct. It was already well known that French, Italian, Spanish, and so on, were "dialects" descended from Latin; William Jones suggested that these formed but a branch of a much larger genealogy. That is, he suggested that most of the European, Iranian, and Indian languages—from those of ancient times to those currently spoken—belonged to a single extended family. In 1814 the British scholar Thomas Young labeled this family Indo-European, a name that eventually displaced other contenders: *Indo-Germanic, Japhetic,* and a tenacious variant, *Aryan.*[3]

Continental scholars followed up Jones's insight, inventing, along the way, the field of comparative philology. A disciplinary hybrid, comparative philology began its career working with the same humanistic materials examined by traditional classical philology: the inscribed word in ancient tongues. All the same, it sought to conduct its investigations along strictly "scientific" lines; it applied itself not to the interpretation of literary content but to the minute dissection of language itself—the anatomy of words and the analysis of grammar. Friedrich Schlegel's *Über die Sprache und Weisheit der Indier* (1808) did more than any other text to set the new field within this natural-scientific framework. As the title of his treatise suggests, Schlegel was concerned with the *Sprache* and *Weisheit,* the language and wisdom, mainly of the ancient Hindus. Yet he contributed something else as well by setting forth a blueprint for the comparative study of all Indo-European tongues. Schlegel acknowledged his debt to William Jones and other British Orientalists, yet he gave his program an original name, "comparative grammar," and declared that this mode of study would illuminate the genealogy of languages "in a similar way as comparative anatomy has illuminated the higher natural history."[4] Schlegel alluded here to the French naturalist Georges Cuvier's work in comparative anatomy and embryology; more broadly, he attached philology to the historicizing of nature which had begun in the second half of the eighteenth century, following the appearance of the Comte de Buffon's *Histoire naturelle* (1749).[5]

Schlegel's *Sprache und Weisheit* constituted a significant piece of intellectual promotional work. More specifically, it served to delineate comparative philology's scope. It suggested that the field should concentrate on languages in isolation, independent of the literary traditions or political histories with which they were associated. Schlegel and his followers used the term *comparison* to denote a process whereby the earliest surviving traits of a language were distinguished from those that had entered the tongue more recently. By isolating those features held in common by a number of languages, philologists could establish the most archaic shared forms and thereby discover the "genealogical" lines through which those languages had developed. Linguistic classification was thus determined chiefly by historical pedigree. For example, Gothic and Latin shared more traits and thus were more closely related than were their respective descendants, modern German and Italian.

Friedrich Schlegel's proposals spurred other Continental researchers: the first of these, Franz Bopp (1791–1867), became comparative philology's first real practitioner. Bopp declared it his purpose to give a "more scientific" treatment of language, one that would "trace the natural-historical laws according to which occurred its development or destruction or rebirth from previous ruination."[6] With his *Conjugationssystem* (1816), and especially the first volume of his *Vergleichende Grammatik* (1833) (Comparative grammar), Bopp broke new ground by comparing the verbal inflections of Sanskrit, Greek, Latin, Persian, Germanic, Zend, Armenian, Lithuanian, and Old Latvian.[7] This analysis demonstrated the affiliation of these Indo-European tongues, now subdivided into their Romance, Germanic, Slavic, Celtic, and Indic branches.

An alternate method of comparative linguistic study also arose at this time, one focused not on word structure but on phonetics. If a given sound x in one language were found to be systematically represented by a given sound y appearing in the cognate words of another language, this provided even surer proof of kinship between these tongues than did agreement in grammatical inflection. The Dane Rasmus Rask pioneered this technique, although the German folklorist Jacob Grimm (1785–1863) made it famous. In his monumental *Deutsche Grammatik* (1819–37), Grimm used phonetic correspondences to reveal the systematic "rules" that bound together the Graeco-Latin and Germanic languages and proved their common ancestry. The result, known as Grimm's law, paved the way for an eventual emphasis on phonetic law in general, even more than on parallel inflection, as the key to tracing individual etymologies as well as to establishing the larger networks of linguistic kinship.

Comparative Indo-European philology reached its early maturity shortly after midcentury, with the career of August Schleicher (1821–68). In his most important contribution, Schleicher reconstructed the basic elements of the Indo-European prototongue. Others had already declared this the goal of comparative philology: said the British linguist J. W. Donaldson as early as 1839, "This reproduction of the common mother of our family of languages, by a comparison of the features of all her children, is the great general object to which the efforts of the philologer should be directed."[8] Schleicher was the first to attempt this systematically. Following Bopp, he listed for comparison the cognate forms of various words and grammatical units as they appeared in several languages. At the head of each column, however, he added his reconstruction of the unattested protoform or root from which the others had descended. This he designated with an asterisk, a practice still used today in historical linguistics. These root reconstructions stood as the ultimate fulfillment of the Indo-European linguistic project.

Yet there was another way in which Schleicher completed what the earlier comparativists had begun: he emphasized the metaphor of family trees of languages. He was one of the first linguists to include in his writings actual tree diagrams; with these he represented visually his conception of genealogical descent and relationship among the Indo-European tongues.[9] Although he conceived the family tree image in the course of strictly linguistic research, Schleicher would soon attach to it intellectual currents outside his field. After reading *The Origin of Species* (1859), he pointed out the striking parallelism between linguistic descent and Darwinian evolution; in this way he launched a good deal of the metaphoric discussion of Darwinism carried on in the 1860s.

Linguistic Themes in British Science

Most of the scholarly work I discussed so far was produced in Germany, for it was there that the discipline of comparative philology chiefly developed. Yet my interest lies ultimately in the English-speaking world, and here one finds, if not a great deal of pioneering linguistic scholarship, then at least an equal and in some ways greater emphasis placed on the analogic relations between philology and natural history.[10] This thesis may appear counterintuitive, for German idealism and *Naturwissenschaft* had initially encouraged the view that language provided an apt analog of biological growth and geological change. Yet by the end of the 1830s, the British scientific elite—including those who probably knew little of the pronouncements of Schlegel, Bopp, or Grimm—would receive a thorough exposure to this cross-disciplinary analogizing. They gained this, moreover, from writers within their own ranks and in the context of some of their most earnest philosophical and methodological discussions. Charles Darwin and his peers, especially the scientific men of Cambridge University and the Geological Society of London, in this way became accustomed to images borrowed from linguistic study a good twenty years before the appearance of *The Origin of Species.*

The "new philology" of Bopp and Grimm began seeping into England in the early 1830s, in part through the efforts of Darwin's cousin and brother-in-law, Hensleigh Wedgwood (1803–91). Wedgwood later helped establish the Philological Society of London and prepared the etymologies for the original edition of the *New English Dictionary*. Yet he is especially important for his private role in connection with Darwin. The two men conversed regularly during Darwin's

London years, the period just after Darwin's 1836 return from his voyage on board the *Beagle*; Wedgwood no doubt informed his cousin during this time about contemporary developments in philological research.[11] This, however, was but a part of a larger merging of science and philology.

Oddly enough, one of the most significant influences on the linguistic image in nineteenth-century Britain was Charles Lyell's *Principles of Geology* (1830, 1833), that much admired yet idiosyncratic brief for a "uniformitarian" scientific method. Perhaps more precisely referred to as an actualistic method, this was the assumption that the kinds of natural forces observed at work today were the same kinds that brought about changes in the past.[12] Lyell was fortunate to have among his supporters the astronomer John F. W. Herschel (1792–1871), perhaps the most highly regarded voice in British science at that time. From his observatory in Cape Town, South Africa, in 1836, Herschel penned a long congratulatory epistle to Lyell. The letter was really a set of reflections on the Lyellian methodology, which Herschel saw as an extension of his own theory of scientific *verae causae*. Valid causes of unexplained events could be inferred, Herschel argued, by analogy with forces observed operating in actual experience. He urged, moreover, that this method be applied in the study of human institutions such as language: "Words are to the anthropologist what rolled pebbles are to the geologist—battered relics of past ages often containing within them indelible records capable of intelligent interpretation."[13] That is, words bore marks of their original composition and so could yield insights into the "anthropological" past.

Lyell circulated Herschel's letter among his London colleagues; Darwin was one of its readers, and he, at least, was particularly impressed by the image of linguistic relics. (Later, perhaps mistakenly, he recalled Herschel having conveyed the analogy to him in conversation during the *Beagle's* visit to the Cape of Good Hope.)[14] Soon Darwin was experimenting with his own linguistic analogies, ones with which to illustrate his grand theory of the evolution of species.

The late 1830s saw an increased pace and intensity of linguistic allusions. Herschel's mention of the similarity between word composition and geological conglomerates reappeared a year later in the work of an opponent: the Cambridge mathematician William Whewell's *History of the Inductive Sciences* (1837). Probably more than any other English-language texts prior to the 1860s, Whewell's *History* and his companion volumes on the *Philosophy of the Inductive Sciences* (1840) popularized the notion of a parallelism between natural history and language study. Darwin and his associates read Whewell's works, but just as important, so did a broad lay public: Whewell's *History* and *Philosophy* went through several editions each during the 1840s and 1850s. These works stood alongside Lyell's *Principles* as authoritative statements of Victorian scientific thought, although they too were substantially polemical. A philosophical rationalist, Whewell rejected Herschel's as well as Lyell's versions of inductive reasoning. Moreover, what Herschel saw as only an isolated analogy (words as "rolled pebbles"), Whewell saw as a grand parallelism among a number of intellectual disciplines.

In the final chapters of his *History*, Whewell discussed what he called the "palaetiological" sciences: showing considerable imaginative boldness, he classed geology and paleontology alongside ethnology, archaeology, and comparative philology. These fields treated an array of different phenomena, including the earth, its past inhabitants (both animal and human), and the history of human kinship, culture, and language. Even so, Whewell argued, these studies shared a common methodology, for they all sought to ascertain "a past state of things, by the aid of the evidence of the present."[15] Yet this was not all, for he posited a basic similarity among the substances themselves: each of these fields manifested a developmental sequence that unfolded over time, "the phenomena at each step [becoming] more and more complicated, by involving the results of all that has preceded."[16] Even though Whewell rejected biological transmutationism (he would later assail that doctrine), he described here a fluid process of change, distinctly evolutionary in the broad sense of the term.

Especially striking was Whewell's argument that the unity among the palaetiological sciences, comparative philology included, was a matter of deep structure: "In asserting, with Cuvier, that 'The geologist is an antiquary of a new order,' we do not mask a fanciful and superficial resemblance of employment merely, but a real and philosophical connexion of the principles of investigation."[17] Whewell reiterated this point, again noting the tendency among geological writers to adopt the lexicon of the antiquary. The description of present-day earth formations as "relics and ruins of earlier states" did not arise from a mere literary habit of mind. Rather, "the analogical figures by which we are tempted to express this relation are philosophically just [later editions: "philosophically true"]. It is more than a mere fanciful description, to say that in languages, customs, forms of Society, political institutions, we see a number of formations super-imposed upon one another, each of which is, for the most part, an assemblage of fragments and results of the preceding condition." For an example of this developmental pattern, Whewell borrowed John Herschel's image of words as rolled pebbles.[18]

In its immediate use, this borrowing and the argument behind it served as a rationalist polemic, part of Whewell's effort to show that Herschel's and Lyell's own attraction to language-based imagery stemmed from a deeper resonance than they themselves realized. For Whewell, the palaetiological sciences manifested a true or inherent analogy between the human-antiquarian and natural-historical fields of study, something deeper than Herschel's concept of causal analogy or Lyell's actualism. This debate would have drawn the attention of British geologists and other natural scientists, exposing that audience to a common search for a way to unify these disparate fields conceptually and so promote a consistent scientific procedure. At this high level of generalization, Herschel's and Whewell's positions actually melded together: above their differences in scientific philosophy and geological doctrine hovered the common language, as it were, of disciplinary convergence, including even the imagery in which this was to be expressed. Whewell, Herschel, and Lyell agreed that individual points of comparison between the linguistic and natural-historical fields reflected the nature of those fields, however this was construed philosophically.

For their part, British philologists were eager to second this thesis: both W. B. Winning's *Manual of Comparative Philology* (1838) and J. W. Donaldson's *New Cratylus* (1839), contain comparisons with natural-historical study, especially geology. Said Winning: "It is now as much the business of the Philologist to recover the remoter history of man, through the fragments of dead languages, in the use of Comparative Philology, as it is of the Geologist to unveil the history of former worlds, from the fossil remains of extinct animals by means of Comparative Anatomy." Donaldson likewise stressed the actualist character of philological study, which he said was "indeed perfectly analogous to Geology; they both present us with a set of deposits in a present state of amalgamation which may, by an allowable chain of reasoning, in either case deduce from the *present* the *former* condition."[19]

All of this suggests that Charles Darwin was not acting as an isolated thinker when he came up with analogies to illustrate his species theory. Rather, he participated in a close-knit discursive world, whose shared theoretical concerns and rhetorical usages were already promoting a sense of philology's natural resonance with other scientific fields. A tradition of linguistic analogy making was thus already in place, along with an impressive effort to justify its validity, well before Darwin set forth similar comparisons in print. Further use of such comparisons would build upon existing practice.

2

From the Early Notebooks to
The Origin of Species

WHEN CHARLES DARWIN recorded his earliest observations on language, he made a crucial addition to the natural-historical view of that subject already in place. He introduced the novel element of transmutation. Darwin's interest in linguistic matters developed concurrently with his theorizing about biological evolution, resulting in a number of references to language scattered through his early notebooks, his unpublished manuscripts, and *The Origin of Species* (1859). The first such references appear in 1837, soon after Darwin completed his five-year voyage on board the *Beagle*. During the final months of that journey, he began keeping a series of notebooks in which he would record the next several years' worth of intensive scientific theorizing. Here the patient reader can retrace the mental steps leading up to Darwin's major breakthrough, the idea of the continual modification of species by natural selection.

A Way to Put the Argument

Even in his earliest musings on transmutation, Darwin pushed his theory into the most controversial realm of all: he began to speculate almost immediately about the evolutionary emergence of man. It is hardly surprising, then, that the linguistic references in his notebooks concentrate on ideas about the origin of human speech. Darwin's eclectic reading during the post-*Beagle* years included a substantial amount of material addressing this subject. He read, for instance, William Gardiner's *Music of Nature* (1832), which argued that the first humans had invented language by imitating sounds occurring in nature. He also read the competing theory of the Scotsman James Burnett, Lord Monboddo (1714–99), which suggested that language had emerged from primeval man's involuntary grunts and groans.[1] Darwin duly noted his responses to these and other conjectures about the first glimmerings of articulate speech.

Of particular interest, however, are a handful of Darwin's remarks noteworthy for the way in which they commingle two distinct ideas. These passages start out as speculations regarding the origin of speech, yet they quickly shade into reflections on the purely analogic similitude between languages and species. For instance, after summarizing William Gardiner's imitative theory of language, Darwin added a parenthetical note: "I may put the argument, that many learned men seem to consider there is good evidence in the structure of language, that it was progressively formed. . . . Seeing how simple an explanation it offers of [the] radical diversity of tongues."[2] In these brief comments, Darwin suggested that the variety of human languages bore witness to an interdependency between structure and history: present-day diversity evinced, indeed was explained by, a bygone process of development.

Even more striking, Darwin suggested that this linguistic configuration held useful material for putting across his argument.[3] It is crucial to see that the argument to which Darwin referred was his case for transmutation in general, not his case for the evolutionary emergence of humankind: emphasizing the world's "radical diversity" of languages would be irrelevant to the latter. Hence even at the initial stage of his career as an evolutionist, Darwin was considering analogic means by which he could commend the idea of common descent to a skeptical public.

There are further, although less direct, indications that Darwin found in language an apt reflection of his transmutation theory. One early notebook entry begins: "At least it appears [that] all speculations on the origin of language must presume it originates slowly—if these speculations are utterly valueless, then [my] argument fails—if they have [value], then language was progressive."[4] At this point, the argument under consideration was Darwin's case for human evolution, which required the gradual emergence—not sudden creation—of mental capacities such as the faculty of speech. Building on this theme of gradual modification, Darwin then shifted perspective. "We cannot doubt that language is an altering element, we see words invented—we see their

origin in names of People—Sound of words—argument of original formation—
declension etc. often show traces of origin."[5] That is, words and whole languages constantly change over time yet preserve within themselves vestiges of
their "original formation."

Darwin returned to this thesis a few pages later in remarking on the teach-
ings of the British linguistic philosopher John Horne Tooke (1736–1812). Could
it be, he asked, that Horne Tooke's analysis of abbreviations and other routine
"corruptions" of language revealed a universal pattern of development, "so that
much of EVERY language shows traces of anterior state??"[6] The possibility that
this might be the case apparently called to Darwin's mind the similar pattern he
posited with respect to the organic world. In sum, Darwin found in linguistic
phenomena a drama of gradual, evolutionary change, one in which "traces of
[an] anterior state" and even "traces of origin" could still be found in latter-day
forms. As he sketched these themes, he surely had in mind a wider field of
application than language alone.

Finally, it is likely that Darwin's reading of Horne Tooke not only reinforced
his idea of gradual evolutionary change but also heightened his awareness of
branching descent among languages. In his monumental treatise *The Diversions
of Purley* (1798, 1805), only a few pages after the discussion of linguistic "abbreviations and corruptions," Horne Tooke made this remark: "French, Italian,
Anglo-Saxon, Dutch, German, Danish and Swedish . . . (together with English)
are little more than different dialects of one and the same language."[7] Here in
essence was the genealogical principle, the idea that quite distinct languages
could descend from a common ancestor. Also in 1840, the same year in which
he read Horne Tooke, Darwin read the early volumes of the *Asiatic Researches*,
the journal of the British Orientalists in India, and "chiefly W. Jones work."
Hence he probably would have seen William Jones's pronouncement about
Sanskrit, Greek, and Latin being "sprung from some common source."[8] By this
time, Darwin's eye for telling metaphors and his penchant for imaginative leaps
were already well honed, and such readings must have helped steer him toward
the language-species analogies that would eventually appear in his works.

From this hasty look at Darwin's notebooks, we can see already the particular "unit ideas" that linguistic images would convey in his early writings. Darwin's transmutation theory may be subdivided into several components, including (among others) evolutionary change as such, gradualism, common
descent, and natural selection.[9] All but the last of these would appear in linguistic guise in Darwin's early works; natural selection and other, more elaborate
aspects of his theory would be treated later. The linguistic analog of evolutionary change was the fact that languages undergo steady transformation over the

course of time; gradualism suggested that this process took place incremen-tally. These two findings of contemporary philology were useful to Darwin, for even if the amount of time required for noticeable linguistic change was really minuscule compared with that required for biological transmutation, it still demanded of its students a similar effort of the imagination. Moreover, this gradual kind of change was characterized by an "organic" unfolding, one that always maintained a degree of continuity with the past and preserved, as Darwin put it, "traces of anterior states."

The other main component of Darwin's theory was the idea of descent from a common ancestor, and some of his early linguistic references accordingly suggested this theme.[10] The notion of evolution as a diversifying process first came to Darwin as he reviewed his biogeographic observations made during the voyage of the *Beagle*. It arose particularly from his puzzlement over the phenomenon of varied populations of the same species type, such as those found on the islands of the Galápagos chain. A divergent or branching view of speciation seemed the inescapable inference from Darwin's discovery that these populations had diversified under their isolated conditions, even though all were related to a previously existing form. As historians of science point out, this branching model was a new feature in transmutationist theory, not found prior to Darwin. The French naturalist Jean-Baptiste Lamarck, writing in 1809, and the anonymous author of *Vestiges of the Natural History of Creation* (1844), each had retained a "great chain of being," a progressive evolutionary ladder extending from the simplest and lowest organisms directly up to humankind.[11]

One can therefore see how the new comparative philology—even though it dealt with an entirely different order of phenomena—supplied a fitting parallel to Darwin's branching-evolutionary schema. And in a sense, philology held claim to priority in such thinking. The work of Franz and Bopp and Jacob Grimm presented the only successful intellectual enterprise, prior to 1859, positing both branching development and real evolutionary descent.[12] Even comparative anatomy, embryology, and paleontology—although increasingly discovering a radiating pattern of forms within each biological *taxon*—still depicted those forms as emerging in discrete succession.

The staunch antievolutionist Louis Agassiz made this point when he distin-guished the reliable views set forth in his *Essai sur la classification des poissons* (1844) from those found in *Vestiges of Creation*, which appeared that same year. Although the *Essai* featured a diagram depicting the branching taxonomic relations among various species of fish, Agassiz assured his readers that this picture was static in character. It was intended, he said, "to show that the genealogical development of species is repeatedly interrupted, and that if, in spite of that, each stock gives us indications of regular progression, this filiation is not really the result of a continuous lineage, but of a reiterated manifestation of the order of things 'determined in advance.'"[13] Comparative philology, on the other hand, dealing with the gradually changing phenomenon of language, entailed not only branching radiation but also evolutionary continuity. Conceptually, then, philology actually stood closer to the idea of descent-with-modification than did pre-Darwinian natural history, even including the work of the reputed scientific forerunners of Darwin.[14]

This structural similarity between the new philology and Darwin's transmutation theory has led some writers to suggest that comparative linguistics helped inspire Darwin in the original formulation of his descent idea.[15] This argument necessarily assumes two things. First, it assumes that philology influenced Darwin's unconscious theorizing, for nothing in his notebooks suggests that he knowingly drew upon linguistic ideas. (The allusions to language seen thus far appeared after Darwin formulated his essential theory.) Second, this argument assumes that the tree of life diagrams sketched in Darwin's B Notebook and elsewhere were distinctly philological in character. This could be true, but only as a matter of the most general cultural context, the idea of linguistic genealogy being "in the air."[16] One should be skeptical, then, of the argument for a direct philological influence on Darwin, for biogeographic observations surely provided the most important impetus to his belief in branching descent.

This having been said, one still might expect that images of linguistic divergence would have proved particularly attractive to Darwin as an aid in the presentation of his ideas; indeed, we have already seen how his notebooks paid attention to the "radical diversity of tongues." By thus alluding to philology's branching thesis, Darwin could bring together at once, in a single *Gestalt,* a number of different aspects of his theory. And yet, even though that thesis would figure prominently in the linguistic images appearing in Darwin's unpublished works, half of those in *The Origin of Species* omitted it. These non-branching images were, however, no less dramatic.

The Unpublished Works

Five different versions of the linguistic analogy appeared in Darwin's formal writings up to and including the *Origin.* The first came in his essay of 1844, an unpublished synopsis of his theory. The relevant passage dealt with so-called rudiments: structures, such as the wings of the ostrich, that in no way helped

their possessor adapt to its life conditions.[17] These Darwin regarded as hereditary features, albeit ones rendered useless over time through changes in the
environment and the influence of natural selection. That is, he saw them as best
explained by his theory of common descent.

To illustrate this naturally occurring phenomenon, Darwin pointed to the
way in which English spelling had often failed to keep pace with phonetic
changes in the spoken language: "In the same manner as during changes of
pronunciation certain letters in a word may become useless in pronouncing it,
but yet may aid us in searching for its derivation, so we can see that rudimentary organs, no longer useful to the individual, may be of high importance in
ascertaining its descent, that is, its true classification in the natural system."[18]
An example on the linguistic side would be the "silent" letters in the spelling of
light, which betray that word's kinship with the German *Licht* and so suggest
the descent of English from Germanic stock. Whether orthographic or biological, rudiments were therefore useful anachronisms, bearing witness to earlier,
genetically related forms. In this immediate context, moreover, Darwin argued
that genealogical descent laid the foundation of the true or natural system of
biological classification. He thereby placed the phenomenon of rudiments
within the larger framework offered by his vision of genealogical taxonomy.
Hence, in one of the earliest systematic expositions of his theory, Darwin at
least implicitly paralleled the "Tree of Life" with the family tree of language.

Darwin's second such image was in some ways his most striking, for it
graphically highlighted the linguistic tree's branching profile. This appeared in
his "species book," written during the years 1856–58 yet never completed or
published in its entirety. Darwin had envisaged *The Origin of Species* as but a
preliminary abstract of this work; hence much that would appear in the *Origin*
is anticipated here.[19] For instance, Darwin acknowledged the apparent unlikeli-
hood of widely different species sharing a common ancestor. Why, if several
forms had descended from a common stock, did there not exist "innumerable
varieties or the finest links connecting in an unbroken chain such species?"[20]
The solution to this difficulty, he argued, lay in a right understanding of the
temporal dimension, the idea that present-day gaps between organic forms
were the result of divergence over time. According to Darwin's theory, many of
the species formerly comprising large genera would have been swept away
through extinction, leaving gaps between each of those genera's most divergent forms. Intermediate links therefore would have existed only in the distant
past, so that, in the natural course of development, entire *taxa* would have
stood distinct from one another for long ages. It was unfair, then, to demand
present-day connective links as proof of common descent.

Darwin probably got his idea for illustrating this argument from reading Herbert Spencer's "Progress: Its Law and Cause" in the April 1857 *Westminster Review*. The philosopher Spencer already was arguing for a kind of organic evolution as part of his cosmic law of progress. And among the various phenomena manifesting this general law, he included the "divergence of words having common origins." Through a process of gradual ramification from a primitive root, Spencer said, "there is finally developed a tribe of words so heterogeneous in sound and meaning, that to the uninitiated it seems incredible they should be nearly related."[21]

With this idea in mind, Darwin asked his brother-in-law, Hensleigh Wedgwood to recommend examples of markedly dissimilar words linked historically to a single prototype. Wedgwood had plenty of such material on hand, for he was preparing etymologies at this time for the *New English Dictionary*. He suggested using two related words from the same language, such as *head* and *chief*. The filiation of these words stood out plainly when the historical connections between them were arranged in a graduated series: *head, heved* (Old English), *heafod* (Anglo-Saxon), *haupt* (German), *haubith* (Gothic), *caput* (Latin), *capo* (Italian), *chef* (French), and *chief*. On the other hand, Wedgwood noted, "if we had only E, It & Fr [English, Italian, and French] remaining nobody would have guessed it possible that head & chief could be different forms of the same word. Perhaps one or two striking instances as this & bishop afford a better illustration than a longer series of less decisive ones."[22]

Wedgwood's proposed illustration conveyed brilliantly the logic of divergent descent. Even so, Darwin did not use it. Instead of *head* and *chief,* he preferred his cousin's other suggestion, mentioned only in passing. The English word *bishop* and its French equivalent, *évêque,* were both derived from the Greek *episkopos.* Darwin included this etymology in his species book, at the end of a chapter addressing the difficulties in explaining "Passages from Form to Form." Here he admitted how easy it would be for skeptics to make his theory appear ridiculous: one needed only to ask whether the rhinoceros and gazelle, the elephant and mouse, the frog and fish, each could have descended from a common progenitor. "Involuntarily one immediately looks out for a chain of animals directly connecting these extreme forms." Then came the analogy: "The case is almost parallel with that often encountered by philologists: to one who knew no other language, dead or living, besides French and English, how absurd would the assertion seem, that *évêque* and *bishop* had both certainly descended from a common source, and could still be connected by intermediate links, with the extinct word 'episcopus.' "[23]

It is interesting to see how Darwin used this analogy in the context of the overall presentation of his argument. The heuristic potential of the comparison was great, for it effortlessly represented two dimensions at once: it showed an absence of intermediate forms linking contemporaneous related phenomena, and it supplied the diachronic explanation of those gaps between forms.[24] Yet in spite of this apparently perfect analogic fit, Darwin put the *bishop* illustration to only modest use. It could have illuminated his argument at a number of levels, yet he used it only to dispel the most obtuse objections: "The utmost which I wish, is to deprecate mere ridicule."[25]

More importantly, Darwin chose not to include this illustration in *The Origin of Species;* the entire surrounding paragraph is omitted from the conclusion to chapter 6, "Difficulties on [the] Theory." He did invoke the idea of divergent word forms later, in passing and without referring to the *bishop* example itself, in the introduction to his *Variation of Animals and Plants under Domestication* (1868).[26] Yet it still seems mysterious that Darwin did not deploy an analogy of this kind to greater effect in one of several possible spots in the *Origin*. The explanation perhaps lies in his not wanting to overdo the appeal to language, for he introduced in that book several new versions of the image. One of these, moreover, although not as vivid as the *bishop* example, would be put to especially strategic use. Having cited *bishop / évêque* to deflect "mere ridicule," Darwin would call upon this new analogy to address what he described as "the gravest objection" that could be raised against his theory.

Analogies in the Origin

The three remaining linguistic figures in Darwin's early writings all appeared in *The Origin of Species*. (A fourth and final example in that book was the same as the one in his essay of 1844, a comparison of biological rudiments with archaic spellings.)[27] The first of the *Origin*'s analogies expressed the idea of continuous and gradual change, yet not the notion of branching. It reinforced the fundamental evolutionary thesis that organic species were not fixed and discrete but were historically linked through mutation. The comparison appeared near the end of the book's first chapter, "Variation Under Domestication," itself an extended analogy between natural selection and the arts of domestic animal and plant breeding. This larger analogy focused attention on a kind of biological change familiar to many Victorian gentlemen yet often betraying an element of mystery. As Darwin pointed out,

> a breed, like a dialect of a language, can hardly be said to have had a definite origin. A man preserves and breeds from an individual with some slight deviation of

structure, or takes more care than usual in matching his best animals and thus improves them, and the improved individuals slowly spread in the immediate neighborhood. But as yet they will hardly have a distinct name. . . . When further improved by the same slow and gradual process, they will spread more widely, and will get recognized as something distinct.[28]

Like a new dialect, a new breed of an animal species usually has no clear-cut beginning but emerges imperceptibly from a preexisting stock. This analogy suggested only unilinear, not divergent, modification, yet Darwin used it to buttress one of the first points he wished to establish in presenting his theory to the world. That point, evolutionism as such, played a logically more fundamental role in the Darwinian scheme than did the idea of branching descent. Hence the primary value of the linguistic image derived from the fact that language was never settled but always in flux, suggesting a deep logic of anti-essentialism.[29] Here again, Darwin probably took inspiration from Herbert Spencer's article on progress: in a series of examples, Spencer noted the similar developmental transformations undergone by human languages, domestic animal breeds, and species in nature.[30]

The second linguistic analogy in Darwin's *Origin* was more complex than the first in both structure and import. In a sense, it replaced the absent *bishop* illustration. It appeared at the conclusion to "The Imperfection of the Geological Record," a chapter not included in the unfinished species book. The problem here was closely intertwined, nonetheless, with the issue addressed by the *bishop* analogy, the absence of forms intermediate to related present-day species. Because Darwin attributed these synchronic gaps to divergence over time, he was obliged to account for the nonexistence of fossilized remains attesting continuous diachronic links between extinct and living organisms. He conceded that the number of these transitional forms would have had to be "enormous." "Why then," he asked, "is not every geological formation and every stratum full of such intermediate links? Geology assuredly does not reveal any such finely graduated organic chain; and this, perhaps, is the most obvious and gravest objection which can be urged against my theory."[31]

Darwin's solution drew upon Charles Lyell's landmark *Principles of Geology* (1830–33) as well as his own field research conducted during the voyage of the *Beagle*. Darwin endorsed Lyell's view that fossil-bearing rock was formed only under conditions of subsidence. Since fossils were created only under this limited condition, and since the life span of a species must be quite long relative to the life span of any geological formation, one should not expect the fossil record to reflect the complete history of life. Therefore, one could not infer from the lack of remains of past species that such species had never existed. On this much Darwin and Lyell agreed. Harnessing this argument to his own special purpose, however, Darwin concluded that paleontology presented no contradiction of his belief that many forms "assuredly have connected all the past and present species of the same group into one long and branching chain of life."[32]

Capping off this discussion, Darwin appealed to what he called "Lyell's metaphor." He constructed an elaborate fictive case comparing the fossil record to an intermittently written chronicle of human history, penned in a "slowly changing dialect." Only the final volume had survived, and this treated only the most recent two or three centuries. "Of this volume, only here and there a short chapter has been preserved; and of each page, only here and there a few lines." Hence the "same" word in the evolving dialect would assume a substantially different shape in each of the chapters in which it happened to appear. These words represented the "apparently abruptly changed forms of life, entombed in our consecutive, but widely separated [geological] formations." In light of this hypothetical case, Darwin concluded, the difficulties for his theory presented by the fossil record were "greatly diminished, or even disappear."[33]

Although cumbersome, this illustration was actually simpler than the *bishop/évêque* analogy: it suggested unilinear rather than branching development and it dispensed with the use of any real philological data. Even so, this Lyellian image also held an advantage over the *bishop* analogy, perhaps explaining why Darwin preferred the one over the other in the published "abstract" of his theory. The *Origin's* analogy emphasized a meagerness of historical evidence, suggesting that the gaps in the diachronic chain of life were only apparent rather than real. In the *bishop* illustration, the gaps between contemporaneous words/species had been real indeed, the natural result of branching descent. That vivid instantiation of the branching effect had of course been the *bishop* image's strong suit.

Yet Lyell's metaphor achieved something perhaps even more important, for it tacitly alluded to the similarity of the challenge that confronted philologists and biological transmutationists: they both needed to demonstrate a continuous gradation linking certain present and past phenomena in the absence of complete attestation to that effect.[34] Because of its growing reputation as a successful scholarly discipline, comparative philology tended to normalize this use of incomplete evidence. It had convincingly demonstrated that no present-day language had sprung up new but that each had come down from some extinct progenitor.[35] And it had shown this in spite of the scarcity of written texts surviving from the ancient and medieval eras—a real-life situation not unlike Darwin's imaginary picture of the poorly maintained chronicle.

Although it alone could hardly have swayed the minds of the *Origin*'s readers, Darwin's appeal to the case of language at least suggested the unreasonableness of demanding copious attestation of transitional, much less primordial, organic forms. Following in Lyell's footsteps, Darwin employed this image according to the principle of "eliminative induction": a scientific theory could not be falsified by the absence of evidence in its favor, especially if one could demonstrate that such absence should be expected in the first place.[36] In short, the linguistic analogy offered reasons for not rejecting the idea of biological descent.

Yet a puzzle still remains: whence came the inspiration for Lyell's metaphor? It is well known that Darwin read *Principles of Geology* while on board the *Beagle*, that Charles Lyell thereafter became a close friend and mentor, and that Darwin greatly admired the methodological spirit of Lyell's science. It is equally well known that Darwin overturned Lyell's arguments for a steady-state view of natural history and, especially, for the fixity of species. Yet little has been said about the way in which this double-edged response included Darwin's appropriation of Lyell's illustrative material. Lyell's metaphor, as it appeared in *The Origin of Species*, did not correspond precisely to any single passage in Lyell's writings. Rather, it combined features from two separate illustrations. Darwin alluded in part to Lyell's *Elements of Geology* (1838), to an illustration comparing the geological record to a stack of old history books, each of which was missing many of its pages.[37] Darwin had registered his approval of this image in an early notebook entry, made in the year in which the *Elements* appeared. There he observed how easily the analogy might be extended to address his own problem: "Lyell's excellent view of geology, of each formation being merely a page torn out of a history, and the geologist being obliged to fill in the gaps—[it] is possibly the same with the Zoologist who has [to] trace the structure of animals and plants—he get[s] merely a few pages."[38]

Yet the metaphor in Darwin's *Origin* contained an additional and crucial feature, a "slowly changing dialect." The likely inspiration for this was a passage from *Principles of Geology*. In this instance, however, Darwin undercut Lyell's argument. *Principles* made a case for a cyclical, nonprogressive view of long-term geological and biological change; it proposed an even distribution, in time and place, of the births and deaths of species. Yet the fossil record hardly reflected this uniform distribution, and Lyell was obliged to make allowance for this by stressing the intermittent deposition of geological strata. He pictured the sedimentation process as visiting a given geographic district only occasionally, thereby leaving an incomplete record.[39] So although there were disjunctures in the fossil record, Lyell argued that these need not imply "catastrophic" interruptions in the smooth course through which species regularly became extinct and were replaced by new creations. On the latter point, Darwin's opinion obviously differed from Lyell's.

Even so, much of Lyell's reasoning was still useful to Darwin. As Michael Bartholomew has argued, the larger historical significance of Lyell's view of the fossil record lay not in its positive content but in the "freedom to speculate" it allowed in regard to geology's missing pages. After declaring that one must expect "to meet occasionally with sudden transitions from one set of organic remains to another," Lyell had cautioned against drawing unwarranted inferences from these clear-cut breaks: "The causes which have given rise to . . . differences in mineral characters have no necessary connexion with those which have produced a change in the species of imbedded plants and animals."[40]

To press this point home, Lyell drew inspiration from his travels in Italy in 1828, during which he had visited the ruins of Herculaneum and Pompeii.[41] In *Principles of Geology*, he imagined two such cities buried at the foot of Vesuvius, albeit one lying atop the other. By unearthing inscriptions from public buildings, an archaeologist might prove that the lower site had spoken Greek and the upper, Italian. "But he would reason very hastily," Lyell warned, "if he also concluded, from these data, that there had been a sudden change from the Greek to the Italian language in Campania." If he later uncovered three buried cities, the middle one having spoken Latin,

> he would then perceive the fallacy of his former opinion, and would begin to suspect that the catastrophes, by which the cities were inhumed, might have no relation whatever to the fluctuations in the language of the inhabitants; and that, as the Roman tongue had evidently intervened between the Greek and Italian, so many other dialects may have been spoken in succession, and the passage from the Greek to the Italian may have been very gradual; some terms growing obsolete, while others were introduced from time to time.[42]

This archeological tableau repays close attention, for a careful reading will highlight, by way of contrast, the subversive nature of the *Origin's* version of Lyell's metaphor. Despite surface appearances, Lyell was not painting a picture of linguistic "descent" of the kind ferreted out by comparative philologists. That is, he was not suggesting the transition from mother to daughter dialects.[43] Since the languages in this illustration clearly stood for biological species, a mother-daughter linguistic relationship would have suggested evolutionary descent among living forms—an idea Lyell rejected through the better part of his career.

Rather, Lyell was alluding to the slow extinction of certain species and their replacement by newly created ones. True, he described a gradual passage from

Greek to Latin and from Latin to Italian.[44] (This last did not indicate modern Italian, but apparently some predecessor dialect, significantly removed from Latin but still old enough to be used in a city that, hypothetically, could have been buried by an eruption of Vesuvius.) Here Lyell probably was thinking of the spectacular Greek papyri found at Herculaneum, evidence of the early Greek settlements in that part of Italy before it became Latinized.[45] This would suggest an external, historically contingent kind of linguistic change, not one of organic growth from within. As Lyell had said, the passage from Greek to Latin to Italian, although gradual, was made up of "many . . . dialects . . . spoken in succession." This transition also involved continuous linguistic intermixture, in which some words grew obsolete while others, borrowed from the newer language of the region, were occasionally introduced.[46] Through this multistage process, the city's inhabitants would eventually acquire an entirely new tongue.

A substantially different story appeared in Darwin's rendering found in *The Origin of Species*. Instead of many tongues "spoken in succession," we saw there a single "slowly changing dialect"—obviously suggesting descent with modification. Still, Darwin retained Lyell's distinction between what could be read firsthand in the earth's strata and what could be known with accuracy about the birth and death of species. He therefore retained the image of a historical chronicle with many missing pages. In one stroke, then, Darwin illustrated how he had both embraced and overturned Lyell's argument concerning the relationship between species and the fossil record.

Yet why was it useful to suggest agreement, to whatever degree, with Lyell? Darwin must have hoped that doing this would help beguile readers into overlooking how far his theory was leading them away from natural-historical orthodoxy. Indeed, he did exactly what Lyell himself later advised Thomas Henry Huxley to do, "to write as if you were not running counter to old ideas."[47] Thoroughly wasted on its intended recipient, this advice captured perfectly the gentle kind of subversion Darwin attempted with Lyell's metaphor.

The book to which he chiefly alluded, *Principles of Geology,* was probably the most popular English-language work on natural science published during the first half of the nineteenth century, and Lyell was still highly respected in the 1850s as a scientific spokesman. Darwin was therefore eager to highlight any resonance, even in illustrative detail, with Lyell's writings. Although a genuinely evolutionary language-species analogy had no precedent, and could have been invented only by a transmutationist, it was nonetheless true that the renowned geologist had produced at least something along these lines in his tale of the buried Italian cities. Moreover, Lyell always placed this illustration in a prominent setting, at the end of an early chapter in the several editions of

Principles.[48] If Darwin could somehow invoke this memorable figure, even while quietly undermining Lyell's essentialist view of species, that would be the best of both worlds.

For this reason, Darwin made his borrowing explicit. With a bit of rhetorical license, he attached Lyell's name to what was really his own linguistic parable— an opportunity obviously not presented by Hensleigh Wedgwood's *bishop* analogy. This tactic of illustrative appropriation actually took place on both sides, for, as I point out in chapter 3, Lyell rebutted in kind: he used linguistic themes to defend yet at the same time to qualify Darwin's evolution theory. Darwin later responded with yet another double-edged move vis-à-vis Lyell, this appearing in his book *The Descent of Man* (1871). Repeatedly, then, each writer took pains to capture the linguistic analogy for his own polemical purposes.

The Family Tree of Man

Of the three linguistic images introduced for the first time in *The Origin of Species,* the last one was the lengthiest and the one most often discussed by later commentators. It gave the fullest display thus far of the analogy's conceptual range, vividly demonstrating how the present diversity and past development of living forms could be represented at once through a single branching genealogy. This arrangement suggested not only a pattern of development but also a system of biological classification, a system based on common descent. Yet in conveying this point, and unlike all of Darwin's previous analogies, this last one added an ethnological dimension. All before this had been, as it were, free-standing, for they had considered language in isolation, detached from communities of speakers. Now Darwin tied language to the larger question of racial kinship.

Darwin had introduced his notion of genealogical taxonomy in chapter 4 of the *Origin.* There he described "the great Tree of Life," whose "ramifying branches may well represent the classification of all extinct and living species in groups subordinate to groups."[49] That chapter also included the *Origin's* famous pedigree diagram, showing how these horizontally placed groups were linked through their emergence on a vertical/temporal axis (see fig. 2.1). Yet it was not until chapter 13 that Darwin discussed in detail how this "strictly genealogical" arrangement solved the age-old problem of biological classification.[50]

Then Darwin presented the analogy: he compared the benefits bestowed by this genealogical arrangement of living forms with the advantages that would be gained from an ethnological classification of languages. Implicit was the idea

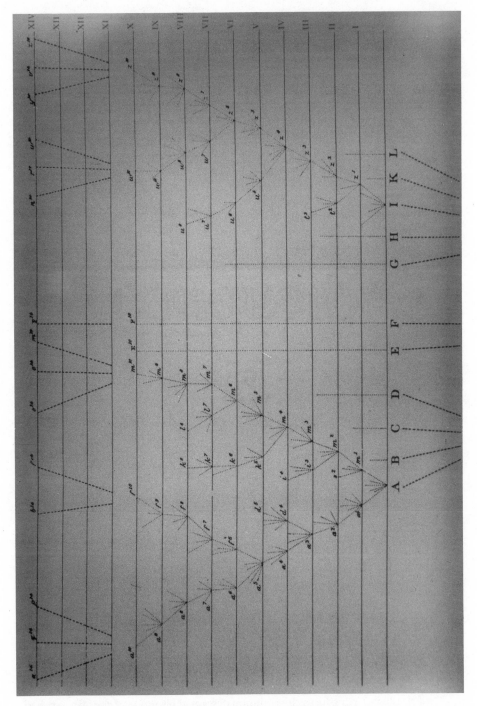

Fig. 2.1. C. Darwin, genealogical chart from *The Origin of Species* (1859).

that languages could be grouped in other possible ways, such as by grammatical type, but that these would not provide as much classificatory data. "If we possessed a perfect pedigree of mankind," Darwin surmised,

> a genealogical arrangement of the races of man would afford the best classification of the various languages now spoken throughout the world; and if all extinct languages, and all intermediate and slowly changing dialects, had to be included, such an arrangement would, I think, be the only possible one. Yet it might be that some very ancient language had altered little, and had given rise to few new languages, whilst others . . . had altered much, and had given rise to many new languages and dialects. The various degrees of difference in the languages from the same stock, would have to be expressed by groups subordinate to groups; but the proper or even only possible arrangement would still be genealogical; and this would be strictly natural, as it would connect together all languages, extinct and modern, by the closest affinities, and would give the filiation and origin of each tongue.[51]

The reader of this passage can easily lose the sense that a comparison between very different phenomena is being made. The reason for this is that the passage itself establishes a comparison between racial and linguistic classification only; the analogy's biological import is a matter to be pondered.

It is harder in the case of this illustration than in those looked at previously to specify its intended lessons, for they were nearly inextricably interwoven. Still, dividing somewhat arbitrarily, one can see this analogy serving three functions. First, it implicitly linked Darwin's species theory to the popular doctrine of racial "monogenesis." In the decades leading up to the *Origin*, the fundamental question of British ethnology concerned racial or ethnic filiation: were humanity's several racial types descended from a single original stock, or were they independent creations, hence their differences primordial? Orthodox religion favored the first of these positions; the Mosaic genealogy of nations in the book of Genesis affirmed what would later become standard church teaching, that all of the earth's peoples were of one blood. This was also the position, although argued on other grounds, of most early-nineteenth-century British ethnologists; it was associated especially with the field's two leading representatives, James Cowles Prichard and Robert Gordon Latham. Implicitly then, Darwin appealed to this monogenetic tradition when he drew his illustration in chapter 13 of the *Origin*. The doctrine of racial "polygenesis," on the other hand, was by analogy antithetical to the Darwinian idea of descent, for it suggested parallel and genetically unconnected lineages.

The second function of the *Origin*'s illustration was to highlight the virtues of classifying living forms by common descent. This mode of classification had been part of ethnological thinking long before it became associated with Darwinian biology, and Darwin himself was keenly aware of the precedent it set. That awareness can be seen in the penciled marginalia he left in his copy of R. G. Latham's *Man and His Migrations* (1851). Here Darwin found *"excellent remarks on classification by descent and resemblance."* (That is to say, descent as opposed to resemblance.) Latham regarded the classification of human groups by physical resemblance—according to skin color, hair type, or skeletal form—as arbitrary and superficial. Such factors, he argued, led researchers to lump together what should be kept separate and to separate what was really closely related. For example, dark-skinned peoples were commonly divided between Fuegans, Eskimos, and "Hottentots" in one group, and African Negroes and Australians in the other. Latham held these categories to be unnatural and indeed valueless if one were asking "how certain areas were peopled; what population gave origin to another; how the [native] Americans reached America; whence the Britons came to England, or any questions connected with the migrations, affiliations, and origin of the varieties of our species." Darwin especially underscored the next remark: admittedly, Latham said, classification by resemblance might reveal certain things, "but not what we want to know, inasmuch as our question now concerns blood, descent, pedigree, relationship."[52] Writing in 1851, Latham could not have known how familiar his words would look when seen through Darwin's eyes. Yet such was the argument that he, Prichard, and others set forth, that their writings seem almost to have preinvented a Darwinian world.

The resonance between ethnology and Darwinian biology was more than a product of imaginative analogizing. The stage had been set by the ethnologists themselves, who already regarded their field as a branch of natural history: trained as physicians, J. C. Prichard and R. G. Latham treated human types according to the model of inquiry into the varieties of animal species.[53] A seeming paradox, however, their classification of peoples according to blood descent depended heavily on linguistic evidence. Its title notwithstanding, the many editions of Prichard's *Researches into the Physical History of Man* (1813) gave considerable attention to the linguistic indicators of racial kinship.

Rooted in an older view of the relationship between language and *ethnos*, this approach regarded language as a kind of natural trait, closely bound up with race, in the traditional, more flexible sense of that term. This point adds a layer of complexity to the picture of linguistic study sketched in chapter 1. There we saw philology itself touted as a kind of natural science, paralleled especially with comparative anatomy. Now we see how ethnology worked a third strand into the weave for, conceptually speaking, zoology, race, and language were all intertwined. Linguistic study did not yet stand as a separate field but was paired with ethnology and, through it, was linked to the biological domain.

The big ethnological questions, concerning the classification of the "nations" of mankind, therefore motivated much of the early nineteenth century's interest in comparative philology. Franz Bopp expressed the common view when he declared that "the genealogy and antiquities of nations can be learned only from the sure testimony of languages themselves." (This epigram appeared on the title page of the British writer W. B. Winning's *Manual of Comparative Philology* [1838].)[54] R. G. Latham put the matter somewhat more carefully: "Whatever may be said against certain over-statements as to constancy, it is an undoubted fact that identity of language is *primâ facie* evidence of identity of [ethnological] origin."[55] Not necessarily a conclusive witness, language nonetheless appeared to offer a more reliable indication of blood kinship than did any other physical or cultural feature. At least through the 1850s, then, most ethnologists hoped that, by drawing up a genealogical pedigree of the world's languages, they could discover more or less the paths by which the divisions of the human family had branched out from a common type.[56]

This backdrop concerning the role of language in the ethnologists' program helps us understand the third and least obvious function of Darwin's genealogical illustration in chapter 13 of the *Origin*. The passage served to vindicate Darwin's strategy of supplying an abstract model, rather than a detailed history, of organic descent.[57] Darwin did not, for example, describe the precise lineages connecting existing mammal forms back to certain extinct mammal types known through fossil remains. He had chosen this approach years earlier, after reading Robert Chambers's notorious *Vestiges of the Natural History of Creation* (1844). That book set forth the most widely discussed theory of biological transmutationism since that of Lamarck, and it was vilified almost universally on either religious or scientific grounds. Darwin watched as reviewers heaped scorn on the anonymous author, in part for his detailed speculations about the ancestries of present-day species. Thus forewarned, Darwin jotted a personal memorandum in his copy of the book: "I will not specify any genealogies— much too little known at present."[58]

The *Origin's* description of biological descent therefore possessed an abstract quality, and this, in turn, was reflected in the distinctly hypothetical tone of chapter 13's analogy. First of all, that ethnological scenario reversed actual practice. Darwin did not say that a complete genealogy of languages would offer the best classification of the races; that would have been an ethnological commonplace. Rather, he said that a genealogical arrangement of the races would afford the best classification of languages; and this would be so only "if" a "perfect pedigree" of the world's races were available in the first place. But then, an independently constructed (and perfect) racial pedigree was hardly to be expected, because of the lack of anatomical evidence dating from the dawn of human existence.[59] Even more elusive, by implied comparison, would be a full reconstruction of the evolutionary ancestries of all living species.

Still, Darwin suggested, even though the actual route of biological descent left precious few traces, the hypothetical model he offered could itself sufficiently approximate that past reality. "It might be," he said, that some language groups had developed more quickly, and hence diversified more, than did others. Here Darwin alluded to the unequal degree of specialized adaptation commonly found among the several branches of a plant or animal *taxon*, a phenomenon he had discussed just prior to introducing the ethnological analogy.[60] He argued that, even if the linguistic tree grew in such an asymmetrical way, this condition still would find its most appropriate reflection in a genealogical system of classification. By accommodating such contingency in linguistic change, Darwin implied, a genealogical schema could reflect similarly irregular development in the biological past.[61] In this indirect way he tried to show that his abstract picture of biological descent was relatively true to life.[62]

We should note, finally, how this deductive approach to the past helps us to locate the naturalist Darwin within yet another tradition of inquiry about human society, not philological or ethnological in this case, but quasi-historical. This was the genre of conjectural history developed by a number of eighteenth-century Scottish philosophers. Darwin had no doubt seen this practice described when, in the early 1840s, he read Dugald Stewart's sketch of the life and thought of Adam Smith (1723–90).[63] Conjectural history sought a theoretical reconstruction of social arrangements in the distant past, a deductive model by which to account for the development of present conditions. Such reconstructions were judged according to how convincingly they explained what could be directly observed, since detailed evidence about prehistoric society was always less than adequate.[64] This conjectural method bore at least a family resemblance to Darwinism's abstract picture of nature's history.

To read so much into this and *The Origin of Species'* other language-based analogies imputes to Darwin considerable painstaking, subtlety, and imaginative sophistication with even the smallest details of his argument's presentation. Yet we should not be surprised to find Darwin occupied with these things, for such was the attention he gave to argumentation in all its forms throughout his career. In particular, he showed a keen appreciation of the use of metaphor for the purpose of endorsing controversial ideas.[65] He had adopted this perspective as early as his species notebooks in the late 1830s, and he continued it well after producing the *Origin*. During the 1860s, a number of writers would try to do Darwin better by elaborating the kind of analogy that he had introduced in his most famous book. Yet Darwin himself would have the last word on the subject, for he eventually would return to these themes with a greater show of imaginative skill than that displayed by all the others.

⤳ 3 ⤶

The Darwinian Circle and
the Post-*Origin* Debates

A FATEFUL COINCIDENCE in the cultural history of science occurred when Friedrich Max Müller delivered his Oxford lecture "Comparative Mythology" just three years before the appearance of Darwin's *Origin of Species* (1859). Despite its title, Müller's address (soon published) dealt as much with linguistics as with mythology proper. This and his later volumes on the "science of language" did more than any other set of writings to popularize the new comparative philology among nineteenth-century English-speaking readers.

Of particular importance, Müller's "Mythology" essay made a case for the one-time existence of a unified Indo-European (what Müller called "Arian" or "Aryan") tongue, from which had derived most of the subsequent languages of Europe, India, and Iran. Müller introduced this argument with a hypothetical case involving the languages descended from classical Latin. Even if all recorded traces of Latin had been lost long ago, "a mere comparison of the six Romance dialects would enable us to say, that at some time there must have been a language from which all these modern dialects derived their origin in common; for without this supposition it would be impossible to account for the facts exhibited by these dialects." In like manner, a comparison of Latin with classical Greek, Gothic, Old Slavonic, Zend, and Sanskrit showed all of these, despite their differences, to be descendants of a common ancestor.[1]

There is no evidence that Charles Darwin had read Max Müller's "Comparative Mythology" before he wrote *The Origin of Species*. But then, he did not need to. As we saw in chapter 2, Darwin had drawn language-species analogies in his early notebooks and manuscripts, most of these penned well before Max Müller arrived on the scene. Yet Müller's writings, appearing so close in time to Darwin's famous book, did influence the subsequent discussion.

By popularizing contemporary linguistic ideas, Müller inspired scientists other than Darwin to consider the similarities between organisms and languages. Most of those scientists were members of Darwin's inner circle, chief among them the geologist Charles Lyell, the botanist Asa Gray, and the zoologist Thomas Henry Huxley. This chapter examines how these and a number of other thinkers variously engaged the linguistic image in the immediate post-*Origin* years. All able writers, these figures knew the importance of adorning one's polemics with vivid rhetorical images. They also knew that once an argument had become identified with a particular image, that bond of association was difficult to break. A useful analogy or metaphor was therefore something to be protected, even fought over.

Most of those who continued to invoke the language-species parallel in the post-*Origin* period used it in the same way as Darwin had done. They saw it as a defensive tactic, to reduce the force of objections to Darwin's evolution theory and so commend it to a skeptical public. They encountered, however, a major antagonist in Darwin's zoologist critic Louis Agassiz. In the period just prior to the *Origin*, Agassiz already was expressing his opposition to evolutionary ideas in indirect terms, via his own quixotic version of the linguistic analogy. The Darwinians' response involved images like those that Darwin himself had used: impressive in their inventiveness, they were nonetheless fairly straightforward comparisons.

Then there was the case of Charles Lyell, who introduced a new, more complex and ambivalent formulation of the analogy. Starting with his writings, the linguistic image served not only to defend Darwin's theory but also to express the intramural disputes among the Darwinians themselves as they debated one another about the philosophical interpretation of that theory. The American Asa Gray played an indispensable role in constructing this in-house discussion. Indeed Darwin did not at first perceive the full implications of the analogy he had set in motion. That is, he did not do so until they were reflected in Lyell's writings and then filtered through Gray's private commentary. Deciphering these linguistic references, both the elaborate and the terse, requires close readings and considerable explication of intellectual context. Two main components of that context demand particular attention. One was the tradition of natural theology, the religious apologetic that so pervaded early Victorian scientific discourse. The other component involved investigations of race and ethnology, a concern at least implicit in most discussions of comparative linguistics.

The Linguistic Analogy and Victorian Ethnology

The British ethnological discourse in the decades before the *Origin* set the essential context for that book's final linguistic illustration, the one describing a genealogical arrangement of the "races of man." Yet there is more to be said about ethnological questions in this period, for, at least in private, Charles Darwin was already considering their analogic bearing on biological transmutation, even without the added linguistic theme. This may be gauged in part from the reading lists Darwin kept in his private notebooks: in addition to books by J. C. Prichard and R. G. Latham, he read works by Robert Knox, Charles Pickering, Charles White, William Lawrence, Samuel Stanhope Smith, and the hyperracialist American polygenists Samuel G. Morton, Josiah C. Nott, and George R. Gliddon.[2] Excepting Knox and the American school, most of these writers reinforced Darwin's belief in the original unity of the various human types, a belief he shared with perhaps the majority of his scientific peers.

Yet Darwin's perspective also departed from the norm at that time, in that his commitment to racial monogenesis was closely bound up with his transmutationism.[3] The connection was not hard to draw. The comparison between human races and the varieties of a species was already standard in nineteenth-century ethnology and was implicit even in the lexicon of zoology: varieties (or breeds) of an animal species were commonly referred to as "races." This was but one aspect of the way pre-1859 ethnological writings anticipated Darwinism. As Darwin himself wrote in his copy of Prichard's *Researches into the Physical History of Mankind,* "How like my book all this will be."[4] (That is, "How like all this my book will be.") Similarly, in Latham's *Man and His Migrations* (1851), Darwin found an "excellent remark" about "how during encroachment, one var. will obliterate intermediate forms."[5] Latham of course referred to competing human types; Darwin used the term *var.—variety—*so as to widen the application.

Equally telling is Darwin's reaction to a very different kind of ethnological work: Samuel G. Morton's polygenist tome *Types of Mankind* (1854). Remarks penciled in the margins of this work show that Darwin did not object to Morton's belief that the several races constituted separate species of mankind. After all, to the transmutationist, species were only well-differentiated varieties

and varieties were species writ small; so the distinction between the two categories was fundamentally obscure. Yet Darwin still considered the human races
to be descended from common stock, and so he held on to the "common belief" in monogenesis. The only difference, he said, was in the name, "whether
to be called species or variations."[6] For Darwin, then, the heart of monogenetic
doctrine was community of descent, not the claim that the races were necessarily members of one human species. Here again he was thinking analogically,
of descent in the larger arena.

   This kind of thinking would have suggested itself almost as a matter of
course to Darwin, primarily because of the shared branching descent thesis but
also because ethnology sought to produce a natural history of mankind. In
other words, analogical reasoning was built into that field already. Yet branching racial differentiation not only offered a deep-structural analogy with organic evolution but was also subject to possible explanation by Darwin's theory.
Indeed Darwin wanted to see his theory of descent by natural selection applied
to the actual origin of racial distinctions: the manuscript of his species book
shows that he had intended to devote an entire chapter to a discussion of the
"races of man."[7]

   Why did Darwin plan to address this subject, one not really essential to his
general transmutation theory? The answer probably lay in Herbert Spencer's
brief for cosmic evolutionism, published in the *Westminster Review*. Writing in
1857, just as Darwin was preparing his intended magnum opus, Spencer declared that his own belief in the branching development of species "gains
immensely in weight" from its harmony with the idea of "divergence of many
races from one race."[8] Even more so, if Darwin's much more detailed ideas of
selection and descent could successfully explain human racial varieties, it
would enhance his overall theory's legitimacy. In the end, therefore, this was
again a kind of race-species analogy, a logical reciprocity between these two
phenomena. Darwin had wanted to include this argument in his big book, yet
he did not have a chance to write that section, and of course did not pursue it
when he prepared *The Origin of Species*. Yet he did address the topic later, in *The
Descent of Man* (1871) (see chap. 4, below).

   We have seen that, during the first half of the nineteenth century, British
ethnologists regarded language as the key to human prehistory, especially the
means to discovering the original racial affinities among the world's peoples.
They based this quest on the assumption that, in general, each language had
adhered over time to its historic tribe of speakers. Languages were in some
cases lost or superseded, yet their tenacity had to be reckoned with as well.
Accordingly, if two unlike physical types were "related" linguistically, J. C.

Prichard tended to assume that those physical differences were not primordial but had been brought about through later environmental influence. R. G. Latham made the complementary point: even though two populations may agree in hair, skin, and so on, without being related, "it is hard to conceive how they could agree in calling the same objects by the same name, without a community of origin, or else either direct or indirect intercourse." This was a circumspect version of the linguistic argument for proof of racial commonality. Latham concluded that, although the investigative usefulness of language had indeed been overrated by some philologists, it had been underrated at times by the more physically oriented ethnologists—the anatomists, archaeologists, and zoologists.[9]

Prichard and especially Latham exercised caution in making this language-centered case for monogenism, and their point of view remained popular through the pre-*Origin* decade. Still, at this very time, the foundations of linguistic ethnology were crumbling. This took place for two reasons. First, scholars such as F. Max Müller argued for the logical separation of race and language. In his "Ethnology v. Phonology," a chapter in Baron C. C. J. Bunsen's *Outlines of the Philosophy of Universal History* (1854), Müller described the appropriate relation between these two sciences as one of "mutual advice and suggestion, but nothing more." His argument was reasonable: "phonological" (linguistic) descent was not a necessary covarient of ethnological descent, "except in ante-historical times, or perhaps at the very dawn of history." In practice, then, "the phonologist should collect his evidence, arrange his classes, divide and combine, as if no Blumenbach had even looked at skulls, as if no Camper had measured facial angles, as if no Owen had examined the basis of a cranium." The verdict, pronounced by Müller and others, was that linguistic evidence could neither strengthen nor disprove the arguments of the physical scientist for or against a common origin of mankind.[10]

Even as Max Müller was launching his critique on logical grounds, paleontological discoveries were reinforcing his point. The most famous of these was made at Brixham cave in 1859, the year of Darwin's *Origin*. Here were found hand-fashioned flint tools alongside fossilized bones of extinct animals; this and other mounting evidence placed humanity's first days (regardless of the question of man's creation or descent) far earlier than the date of roughly 4000 B.C.E. that Bishop Ussher had calculated in the seventeenth century.

The result was a radically extended estimate of the human chronology, which in turn exposed a weakness of the standard ethnological project. That project had relied not only on linguistic evidence but also on the traditional belief in mankind's recent origin. Yet because the written traces of language went back only so far, linguistic research could no longer hope to demonstrate whether humankind possessed a monogenetic heritage. Max Müller was only one among many who warned that researchers could no longer aspire to "trace the convergence towards one common source" of all human tongues. Linguistics was thus rendered incapable of casting light on the question of racial origins.[11] Most important for our purposes, this undermining of language-based monogenism affected the language-based analogy with Darwinian evolution. In this sense, Darwin's close identification of language and race in chapter 13 of the *Origin* reflected an older view of the relationship between those two phenomena, one that was changing even as he wrote.[12]

Already in the pre-*Origin* years, another leading scientist besides Darwin was giving serious attention to the analogy between racial descent and an evolutionary view of species. The Swiss zoologist Louis Agassiz had already earned an enviable reputation in Europe by the time he immigrated to the United States in 1846; there he spent the prime of his career as the founding director of Harvard University's Museum of Comparative Zoology. This was an important turning point, coming just after the appearance of Robert Chambers's *Vestiges of the Natural History of Creation* (1844). Spurred by this event, and well before he knew anything of Darwin's theory, Agassiz took his stand as a vocal opponent of transmutationism.

Long interested in biogeography, the study of the distribution of plants and animals over the earth's surface, Agassiz insisted that each species had not only been created separately but also had been created for the "zoological province" in which it currently resided. Hence similar forms appearing in widely distant regions were different species, separately created; they were not varieties that had spread by migration from a single geographic cradle. For as Agassiz and other naturalists all knew, the diffusion-of-species thesis was potentially radical: although it did not necessarily entail evolutionary descent from a common type, it did suggest a requisite precondition for that notion.[13]

Having migrated himself, Agassiz soon extended his concept of zoological provinces to the human races. He found allies, moreover, among the strongly polygenist American school of physical anthropology. His position was reinforced during lecture tours of the American South during the late 1840s and early 1850s, as he observed enslaved Africans in their straitened circumstances. Such experiences brought Agassiz to the conviction that the races were not only distinct but unequal, a view which naturally gained him favor among the southern aristocracy while causing consternation among his friends in New England.[14] Yet Agassiz had not intended to fan the political passions running so deep in the United States at this time; rather, he was reacting to events in natural science, particularly to the appearance of *Vestiges*. Here again, he un-knowingly set himself in opposition to Darwin, for the views of the two natu-ralists were mirror opposites: Agassiz allowed that humanity's several racial types were all of one species, yet he declared that they were nonetheless separate creations, not descended from a single set of parents or from a single ancestral stock. Racial distinctions were thus "primordial."

Paradoxically, then, Agassiz championed a religiously controversial poly-genism—an apparent denial of the early chapters of Genesis—for the sake of traditional creationist belief. By asserting the original diversity of mankind, he intended to combat a monogenetic logic that would apply not only to races but to species in general. He feared that the biblically orthodox who embraced monogenesis did so in heedless fashion, without considering where that doc-trine was leading them. Monogenism, he warned, "run[s] inevitably into the Lamarckian development theory, so well known in this country through the work entitled 'Vestiges of Creation'; though its premises are generally adopted by those who would shrink from the consequences to which they necessarily lead."[15] Agassiz therefore concluded that, in order to be consistently theistic and affirm creationism, one was obliged to embrace racial polygenism.

This line of argument led Agassiz to confront as well the monogenetic view of language. This he did in each of the several statements he made affirming racial plurality, these appearing in Boston's *Christian Examiner* in 1850, in Sam-uel Morton's *Types of Mankind* (1854), and in Josiah Nott and George Gliddon's notorious *Indigenous Races of the Earth* (1857)—the last two works largely defin-ing the American school of ethnology. Agassiz denied that linguistic affinity proved either community of linguistic origin or diffusion from a common source. Rather, he attributed the common structure of related languages to a similar vocal anatomy and like intellectual powers among the peoples who spoke them.[16] He noted that distinct animal species of the same family pro-duced closely allied sounds, as close in intonation as "the so-called Indo-Germanic languages compared with one another." "The brumming of the bears of Kamtschatka," for example, was "akin to that of the bears of Thibet, of the East Indies," and so on. Yet each kind of bear was rightly considered a distinct species, "who have not any more inherited their voice one from an-other, than [have] the different races of men." Agassiz urged the student of language to study these facts: "If he be not altogether blind to the significance of analogies in nature, he must begin himself to question the reliability of philological evidence as proving genetic derivation."[17]

Critics complained that Agassiz's argument could hardly be taken seriously. The Congregationalist *New Englander*, for instance, found that it stood "in defiance of the conclusions of every philosophical linguist living."[18] Yet this was not entirely true. Agassiz had surely borrowed, albeit selectively, from the most philosophical of Europe's early-nineteenth-century linguists, Wilhelm von Humboldt. Humboldt saw deep structural affinities among the languages of each basic grammatical type—isolating, agglutinating, polysynthetic, and inflective—each of which he saw as representing a different stage of humanity's mental development. This picture roughly paralleled Agassiz's transcendentalist zoology, with its notion that each species represented a distinct expression of the mind of God. Applied to the human races, this outlook dispensed with any physicogenetic basis of affinity and classification. Indeed, it considered affinity by descent and diffusion a crude and animalistic perspective, distracting from the true explanation of the shared intellectual patterns interlacing mind, nature, and language.

This interpretation of linguistic affinity was thus not unique to Agassiz but was common to the larger idealistic tradition in which he had been trained. That tradition made strange alliances, for one of Agassiz's admirers in the American South articulated this same view of language. Writing in 1850, the South Carolina theologian James Warley Miles (1818–75) embraced ethnological polygenism, most immediately as a vindication of race-based slavery. Yet as an intellectual corollary, Miles also discounted the genealogical view of linguistic kinship. The occurrence of the same root word in languages of the same family, he argued, could not "be always explained historically and ethnologically, by derivation from each other; but must often be referred back to the Common Laws of the Language-faculty, producing similar results in different people, according as a Common Human Nature has been subjected to similar conditions."[19] Louis Agassiz produced this same linguistic logic for a different purpose, as he sought to forestall the kind of natural-historical thinking that would soon appear in *The Origin of Species.*[20]

Even so, in the years before Darwin's book appeared, Agassiz's views on ethnogeography drew criticism on scientific grounds. The paleontologist Richard Owen addressed the subject in his 1858 presidential address before the British Association for the Advancement of Science. The occasion took place less than three months after the presentation of papers by Darwin and Alfred Russell Wallace, jointly outlining the idea of speciation via natural selection. Owen followed the tradition in which the association's president reviewed the previous year's scientific developments, and his contrasting treatment of Darwin and Agassiz is revealing. Owen responded to the new transmutationist theory with relative equanimity, advising caution and calling for more evidence. Yet he thoroughly rejected Agassiz's thesis that human racial types were indigenous to distinct regions. Owen thereby hinted at the analogic tension between Darwin and Agassiz and implied that Agassiz had committed the greater scientific offense.[21]

Lyell's Reconciliation with Darwinism

Charles Lyell sat in the audience during Richard Owen's British Association address, and he no doubt followed closely the remarks about Darwin and Agassiz.[22] Spurred by his private conversations with Darwin, Lyell had taken an increased interest in the species question in recent years: this showed in a journal on that subject he had kept since 1855. That next spring, when Darwin explained his transmutation theory to Lyell more fully, Lyell urged him to publish it immediately."[23]

For his own part, however, Lyell held back. He would never become more than a reluctant evolutionist, and he wrestled especially with the theory's implications concerning man. He had presented an elaborate critique of Lamarck's views in his *Principles of Geology* (1831–33), and in 1858 he could still write in his journal: "Transmutation is a mere guess at present." True, Lyell conceded by then that the past thirty years had shown Lamarck's position to be "more reasonable [and] less visionary, because no definition has been found of a species." This fact made him "unwilling to dogmatize with the confidence I once felt." Yet Lyell did not want to dogmatize in the other direction either. Even in the spring of 1860, after *The Origin of Species* had appeared, he described that book's contents as "still only the most probable hypothesis as yet advanced."[24]

At this same time, however, Lyell was reaching a modus vivendi with Darwin's theory by giving it a theistic interpretation. Especially remarkable is the way many of the passages in his scientific journal in which Lyell developed this interpretation were cast in linguistic analogies. Lyell used the case of language both to highlight the merits of Darwin's theory and to illustrate the solutions he proposed to the moral and intellectual problems it raised. These passages reveal once again Lyell's inventiveness as an analogic thinker; they also present a rough draft of themes that would arise in public debate over the philosophical import of Darwinism.

Perhaps surprisingly, Lyell made little mention in his journal of the *Origin*'s linguistic analogies and, in particular, said nothing about what Darwin had called Lyell's metaphor. He nevertheless pursued these same kinds of comparison, drawing material from his own reading of Max Müller, Baron Bunsen, and Wilhelm von Humboldt.[25] As he told a friend in 1860, Müller's "Comparative Mythology" lecture "appears to me in the philological part very excellent. The argument for the existence of some aboriginal language, whether it be called Arian or by any other name, seems conclusive, and it must go a far way back, as they branched off into such distant and ancient nations."[26] With this Indo-European idea in mind, Lyell set down in his journal a succession of pro-Darwinian analogies: the number of languages that have existed through time must have been large, yet each must have had a relatively short life span; there was once a period when no language now spoken existed, whereas "all that existed then are now extinct"; linguistic extinction, moreover, was usually a matter of evolution into an entirely new tongue. "Unity and Continuity," Lyell said, did not necessarily imply "a discoverable relationship in every true existing language and it may be so perhaps in regard to some living species, which at times indefinitely remote, sprang from a common progenitor."[27] All of this suggested the plausibility of transmutationist patterns in the biological realm.

Lyell also noted the logical buttressing that racial monogenesis accorded Darwinism. His remarks were doubtless aimed at Agassiz, whose "Essay on Classification" (1857) he had studied: "It would be better to deny the whole filiation of other languages [than] to invent some arbitrary law by which nations independently of each other, and by similarity of mental constitution struck out [i.e., fashioned], and by identity of organs of speech, hit on, similar expressions and sounds." This invented scenario—the one actually concocted by Agassiz—"would be an unphilosophical proceeding." Yet Lyell found Agassiz to be at least "logical and consistent" in denying racial unity. For unless they could be applied as well to the human family, all of the arguments for the common descent of species "fall to the ground."[28]

Despite the sincerity and imagination that Lyell invested in these passages, the pages of his journal show him giving much more attention to a countervailing use of the linguistic analogy. He was brought to this by the monumental question of humanity's status in relation to the lower animals: could natural laws, he asked, "evolve the rational out of the irrational[?]" Nothing that had existed prior to humanity's appearance anticipated the moral realm: responsibility, sentiment, goodness—these were something new.[29] The point, of course, was that evolutionary theories were unable to account for these human qualities.

Lyell quickly reached an accommodation, however, through a theistic recasting of Darwin's argument: "This theory, instead of being mere naturalism, instead of banishing and distancing the supreme intelligence from the work of creation, instead of denying the intellectual causation of all the phenomena of change, instead of separating the mental and natural phenomena, requires a constant increasing intensity of miraculous power . . . [and] a larger portion of the divine intelligence visiting the earth in a mental form."[30] This was, in a sense, a strikingly un-Lyellian thesis, for it went against the assumption that natural forces previously in operation were essentially no different and no stronger than those in effect today. Lyell now reasoned that, in order to explain advancing complexity over time, one must posit an ongoing and intensifying power at work: "If we believe that the originally created types were less complex and advanced than those now flourishing, if we adopt any law of progress, we immediately require a greater intensity of the intervention or manifestation of the First Cause as Time rolls on."[31]

Also, as Lyell said, along with this progressive causal intensity went increasing manifestations of divine rationality. Here Lyell betrayed his reading of the Oxford mathematician Baden Powell as well as the paleontologist Richard Owen—the speaker at the British Association meeting.[32] These writers presented a transcendentalist alternative to the dominant version of British natural theology, for they stressed nature's intellectual plan rather than the adaptive functions emphasized by William Paley and the writers of the famous *Bridgewater Treatises* of the 1830s.

Especially relevant was the work of Owen, Victorian England's leading specialist on the natural history of vertebrates. Owen's theory of organic progression, what he called "conformity to type," posited the introduction of species in graded succession yet all based on the same essential plan. Having read Owen's classic statement *On the Nature of Limbs* (1849), Lyell was familiar with the idea that "the law that the [vertebrate] Archetype is progressively departed from as the organization is more and more modified in adaptation to higher and more varied powers and actions." The end result was "the progressively superadded structures and perfections in higher reptiles and in mammals."[33] Lyell suggested a similar ordering of nature, only endowing the guiding archetype with growing manifestations of creative power: he conjectured "a series of miracles—the supernatural continually intensified, the origination of New Causes acting in conformity with pre-existing laws."[34]

The bulk of the linguistic images in Lyell's journal reflect this theistic argument, weaving together into a complex tapestry his solution to the problem of "man's place in nature." The culmination came on 22 May 1860, the day on which appeared the longest of Lyell's notebook entries outlining theistic progressionism. Lyell saw the long ascent of language as reflecting the progressive career of human reason and moral sensitivity; increasing conceptual refinement in the invention of words was bound up with more and more exalted thoughts.[35] This was really a combined argument, a description of how theistic evolutionism could account for the actual emergence of human mental faculties, and an analogy for the increasing complexity in species over time. Both the real and the analogical aspects of this argument were important to Lyell: concerned foremost with the status of humanity, he had much to say about the intelligent design of language itself. The concern here, however, is with the way Lyell saw the career of language reflecting the increasing embodiment of divine intelligence through the course of organic history.

This analogy clearly cast doubt on the explanatory efficacy of natural selection. Even if transmutation were reasonably demonstrated to have formed the hidden bond connecting the taxonomic levels of organic life, Lyell was yet convinced that "the creation at successive periods of new beings will still be believed in and ascribed to some power wholly exterior to all this mechanism." That is, while the inner workings of the selection mechanism might be discovered and demythologized, its ultimate source was nonetheless metaphysical. Lyell conveyed this point indirectly.

> When we have got rid of the Garden of Eden as a myth and the tower of Babel, and show by what steps new languages grow or divide into dialects or are partially mixed by crossing or deviate into such distinct forms as that their affinities are scarce recognizable, we are still as far as ever from knowing how this marvelous power is produced—this wonderful weapon of thought. The miracle performed in our own time is greater than we could ever have witnessed at any remote epoch to which our science carries us back.[36]

In this and similar passages, Lyell wove in and out of the analogic mode, rarely distinguishing between references to language itself and figurative references to biological species. But then, the blurring of that line was essential to the vividness of his argument.

As these pages from his journal reveal, the geologist Lyell devoted a surprising amount of attention to linguistic themes during the first half of 1860. More than that, he placed the linguistic analogy front and center as he formulated his ambivalent stance toward *The Origin of Species*. That comparison suggested both his tentative conversion to transmutationism and his coming to terms with its moral and theological implications. These two faces of Lyell's response to Darwinism, both acquiescence and reservation, would eventually reach the public in his *Geological Evidences of the Antiquity of Man* (1863). Meanwhile, Lyell was curious about the reaction to Darwin's book among naturalists in America, especially the two leading figures at Harvard, Louis Agassiz and Asa Gray.[37]

Growth and Fragmentation

In Boston, public debate about the Darwinian question began in the spring of 1860, at specially convened sessions of the American Academy of Arts and Sciences. Asa Gray stood for the defense on these occasions, while Agassiz led the opposition. Gray soon published his case in a number of articles, including a series in the *Atlantic* magazine in which he employed a linguistic analogy for the first and last time. He applied the comparison to the problem of biogeography, arguing for the dispersion of each related group of species from some common center of origin. Everyone knew that closely related species tended to cluster in the same regions; most naturalists therefore supposed that lands now "widely disjoined" must have once been contiguous if they supported similar animal inhabitants. They concluded this, said Gray, "just as philologists infer former connection of races, and a parent language, to account for generic similarities among existing languages."[38]

Asa Gray's main interest for us, however, lies elsewhere, not in his own use of the analogy but in his exposition of the argument from design in relation to Darwinism. Eventually we shall see how Gray's theistic evolutionism informed his analysis of the linguistic image, which, in turn, affected Darwin's view of the subject. But first it is necessary to follow the events in sequence.

Gray's *Atlantic* series gave a supernaturalist reading of Darwinism, beginning with the assumption that, unless it were proved otherwise, Darwin's theory accommodated the idea of divine superintendance. More specifically, Gray stressed organic evolution's beneficent ends despite the many "purposeless" and wasteful variations strewn in its wake. He admitted that nature's path ran into many dead ends, yet he maintained that the whole was "designed to improve."[39] The real question, he said, was how to conceive of the impelling force behind this happy trend in a way both scientifically valid and theistic. He declared it a matter of indifference whether one believed in a single creative impetus at the beginning of time, in occasional supernatural interventions, or in the constant and immediate action of "the intelligent efficient Cause." In the end, he said, "all three are philosophically compatible with design in Nature."[40]

Gray's theism was thus more flexible than Lyell's, yet it was nonetheless firm, to the extent that Gray doubted Darwin's naturalistic interpretation of selection. He later expressed himself on this point with considerable bluntness in a letter to Darwin: "Of course we believers in real design make the most of your 'frank' and natural terms, 'contrivance, purpose,' etc., and pooh-pooh your endeavors to resolve such contrivances into necessary results of certain physical processes, and make fun of the race between long noses and long nectaries!"[41] Despite the blatancy of this remark, Gray intended it only as a qualification, and he hoped that his articles would win a sympathetic hearing for Darwinism in America.

Yet what about Darwin himself? What did he think of this defense of his theory, one stressing its compatibility with traditional natural theology? What Darwin thought may have been different from what he said, for his response was effusive. After reading Gray's *Atlantic* series, he lavished praise on its author. "By Jove I will tell you what you are, a hybrid, a complex cross of Lawyer, Poet, Naturalist, and Theologian! Was there ever such a monster seen before?" A close reader, Darwin added, "Your many metaphors are inimitably good." Darwin did as well as he talked. He contrived to have Gray's series reprinted in England and approved Gray's suggestion of an eye-catching new supertitle: "Natural Selection not inconsistent with Natural Theology." Lyell added his blessing, judging Gray the ablest spokesman on the evolution question, "both as a naturalist and metaphysician," on either side of the ocean. This judgment reflected Lyell's own restrained embrace of Darwinism and his basic agreement with Gray's theism, for both men rejected the deification of matter and secondary causes.[42]

The fall of 1860, however, saw the beginnings of a split among the Darwinians, as Asa Gray's theistic outlook fell from favor and T. H. Huxley's agnostic view moved toward center stage. Darwin himself was at bottom unconvinced by Gray's design version of his theory: he admitted this even as he had Gray's articles republished in England—thereby betraying the opportunism underlying his endorsement of the American botanist's views. Gray apparently took this rebuff in stride, and the two men continued on occasion to discuss the question of design.[43] Darwin had already begun to prod Gray on the subject at the end of his "By Jove" letter of praise. He noted that his brother-in-law, Hensleigh Wedgwood, was "a very strong Theist, and I put it to him, whether he thought that each time a fly was snapped up by a swallow, its death was designed; and he admitted he did not believe so, only that God ordered general laws and left the result to what may be so far called chance, that there was no design in the death of each individual Fly. Farewell my good Friend. Yours most truly, [etc.]"[44] Later, when Gray responded to this challenge, he did so by referring to the linguistic image.

Most manifestations of the linguistic analogy in the first years after Darwin's *Origin*, however, did not address such philosophical issues. Although ingenious, these analogies were generally uncomplicated in structure and modest in aim. The writer George H. Lewes set forth a classic version of the argument in his

"Studies in Animal Life," appearing in *Cornhill Magazine* in the spring of 1860.[45] "The development of numerous specific forms, widely distinguished from each other, out of one common stock, is not a whit more improbable than the development of numerous distinct languages out of a common parent language, which modern philologists have proved to be indubitably the case. Indeed, there is a very remarkable analogy between philology and zoology in this respect." The point here was to reduce implausibility, the key phrase being "not a whit more improbable." Like Charles Lyell, Lewes drew inspiration from Max Müller's "Comparative Mythology": he quoted extensively from this exposition of the Indo-European principle, showing how the diverse array of Indo-European languages, like the Romance tongues, amounted to mere "varieties of one common type."

Having presented Müller's convincing argument so fully, complete with a table of verb paradigms from French, Portuguese, Spanish, Italian, and Wallachian—this alongside the same verb compared in Sanskrit, Gothic, Latin, and other old Indo-European tongues—Lewes felt confident in reaching a bold conclusion. "In the same way, we are justified in supposing that all the classes of the vertebrate animals point to the existence of some elder type, now extinct, from which they were all developed."[46] This was a new claim, that philology could not only deflect objections but also give positive justification of belief in biological common descent.

Yet Lewes then backed off. His aim, he said, was to keep those who believed in the variability of species from being charged with "absurdities they have not advocated." It was almost as if he had read in Darwin's unpublished species book manuscript the passage about avoiding mere ridicule: the latter was the function Darwin had envisaged for the *bishop* analogy. In the same vein, Lewes deemed it unfair to ask whether Darwinians taught that a goose had developed from an oyster or a rhinoceros from a mouse and thus try to "render the doctrine ridiculous." Rather, he urged, one ought to find out what "solid argument" Darwin and his followers offered.[47] Although Lewes's use of the analogy hinted at something bolder than anything Darwin had included in the *Origin*, it led in the end to this essentially modest plea.

A similar argument appeared in T. H. Huxley's lectures for workingmen, delivered in 1862. Like Darwin, Huxley found the phenomenon of vestigial biological forms inexplicable if one assumed a hypothesis of special creation. He then added: "In the language that we speak in England, and in the language of the Greeks, there are identical verbal roots, or elements entering into the composition of words. That fact remains unintelligible so long as we suppose English and Greek to be independently created tongues; but when it is shown that both languages are descended from one original, the Sanscrit, we give an explanation for that resemblance." In short, genealogy accounted for similarity. "In the same way," Huxley declared, parallel anatomical structures found in widely different animals gave "striking evidence in favour of the descent of those animals from a common original."[48] Huxley in this way patterned his use of the language analogy strictly according to the *Origin*'s discussion of biological rudiments, only substituting lexical cognates for silent letters as evidence of common descent. As before, moreover, although the parallelism itself was impressive, its import was modest: linguistic patterns demonstrated how the positing of common descent solved certain problems that would otherwise remain baffling.

Even a hostile source drew a comparison between languages and species at this time, an event of special significance for the private commentary it inspired. In Boston, the American Academy of Arts and Sciences' sessions devoted to the Darwinian question ended in the summer of 1860. A regular meeting the next January, however, brought a parting shot from Louis Agassiz. The evening began innocently enough, with a paper on Sophocles and then one on the optative mode of Greek verbs, the latter given by the classicist William W. Goodwin (1831–1912) of Harvard. Goodwin considered whether anticipations of the Greek modal system might have appeared in the "parent language" of the Indo-European family. Citing the work of German philologists, he affirmed the possibility of reconstructing extinct grammatical forms and concluded that Greek and Latin had indeed "derived the rudiments of their modal forms from a common ancestor."[49] Harvard's president, Cornelius Conway Felton (1804–62), endorsed Goodwin's analysis and added his opinion—consonant with scripture—that all of the earth's languages, not just the Indo-European tongues, had descended from a single original.

At this point, Agassiz rose to speak. Contra Goodwin and Felton, the zoologist reaffirmed what he had said in writing during the past decade. He expressed disbelief in the derivation of later languages from earlier ones: like the races that spoke them, each language was "substantially primordial." What accounted for the resemblances between "related" tongues? Again, Agassiz declared that these arose not from community of descent but from a congruence of mental structure among the nations involved. C. C. Felton responded by pointing out a number of similarities of vocabulary and grammar among affiliated languages: these, he maintained, were "utterly inexplicable" on any ground other than that of common derivation.[50] In pressing this point the classicist Felton surely had no wish to take part in the most explosive scientific debate of the century, much less to support the Darwinian side. Apparently, Felton took Agassiz's linguistic pronouncements at face value only.[51]

But the point was hardly lost on Asa Gray, who was present at the meeting. Gray sent Darwin a gleeful account of the academy incident, describing how Agassiz's quixotic opposition played right into the Darwinians' hands. Apparently, the two Harvard colleagues had discussed the issue on previous occasions. Said Gray: "Agassiz (foolish man) admits that the derivation of languages and that of Species or forms stand on the same foundation, and that he must allow the latter if he allows the former—which I tell him is perfectly logical."[52]

Delighted with Gray's story, Darwin repeated it to his friend Joseph Dalton Hooker (1817–1911), director of Kew Botanical Gardens. Agassiz, he said, had "gone almost demented" in "maintaining that Greek, Latin, and Sanscrit are not affiliated but, like the races of men, are autochthonous [i.e., each indigenous to a particular region]! It is impossible to argue better for us." Darwin also passed the story on to Lyell, again declaring that Agassiz's argument was "really favourable in high degree to us." After quoting from Gray's account, he added, "Is this not marvelous?"[53] Lyell, however, was fully aware of the logical bind in which Agassiz had placed himself, for it was the same problem he had seen already in Agassiz's biogeographical theory of racial polygenism.

In this episode, the original Darwinian circle—Darwin, Lyell, Hooker, and Gray—could close ranks and share a laugh at the outsider Agassiz; Darwin spoke naturally of "us" in his letters. This unity, however, was only a surface matter. Although Lyell no doubt agreed about the intellectual hopelessness of Agassiz's linguistic argument, he envisaged his own alternative rendering of the analogy. If this was not to be as overtly hostile to Darwin's theory, it would at least be able to qualify it considerably. Hence the emerging split within the Darwinian circle found its reflection in the growing complexity of the linguistic analogy, eventually taking that image beyond the relatively simple pro-Darwinian messages seen so far. The analogy would then become, at least in part, a contested matter among the Darwinians themselves.

Max Müller's Equivocal Roles

To a remarkable extent, early Darwinian discourse would be influenced by the writings of the Oxford linguistic savant F. Max Müller. Known among scholars for his pioneering work in comparative mythology, Müller gained a following among the general public with his lecture series on the "science of language," given at London's Royal Institute in 1861. In published form, these eloquent lectures attracted a wide readership in Britain and North America. They also introduced a new topic into the debate over Darwinism, for, in the final chapter, Max Müller set forth his views on the origin of speech. As everyone knew, it was the question of human transcendence that gave this linguistic issue its special relevance.

Even though Darwin had said nothing in print on the subject of language's origins and virtually nothing about the related question of human evolution, Müller launched a preemptive strike: he attacked the two leading non-supernaturalistic views of language. First he rejected what he dubbed the "bow-wow" theory, arguing that, if this onomatopoetic principle had applied anywhere, it would have done so in the formation of the names of animals. As it was, extended etymologies gave little evidence of imitative origins in Indo-European roots: "we listen in vain for any similarity between goose [in its root form] and cackling, hen and clucking, duck and quacking." And names like that of the cuckoo, Müller noted, were in a decided minority.[54] In place of both the mimetic and interjectional explanations (Müller designated the latter the "pooh-pooh" theory), Müller proposed a mystical explanation in which nature's essences resonated forth spontaneously in humanity's first root words.

One of the linguistic writers to whom Müller immediately responded was Darwin's cousin Hensleigh Wedgwood. Wedgwood had outlined an imitative theory of the origin of language in the first volume of his *Dictionary of English Etymology,* published in the same year as Darwin's *Origin.* The onomatopoetic principle, he argued, might well constitute a *vera causa,* a force that had been observed in actual operation and which produced new words in a manner comparable on a miniature scale to the first awakening of articulate speech. He declared that geological uniformitarianism had blazed a trail for this kind of scientific reasoning, which awaited its full application to linguistic phenomena.

Wedgwood later pursued that application in his *Origin of Language* of 1866. Yet before doing this, he handed over the reins to his eldest daughter, Frances Julia (1833–1915), who adopted her father's philological interests, and whose literary and scientific mentors included Marian Evans and George Lewes. Julia Wedgwood's unsigned review of Max Müller's origin theory appeared in late 1862 in *Macmillan's Magazine,* the same publication in which Müller's initial lectures had been serialized only a year earlier.[55] Her rebuttal made no direct mention of her uncle Charles's ideas on biological development, yet it strongly suggests that she had this subject in mind. It also suggests that she knew of the special significance of the *bishop* illustration that her father had recommended for Darwin's species book.

Julia Wedgwood mainly addressed the problem that Max Müller had pointed out, that few etymologies appear to point toward imitative beginnings; no quacking, for example, was to be heard even in the oldest known words for *duck*. In response, Wedgwood noted that the phonetic shape of a word can travel far over time from its original root, thereby obscuring all trace of any imitative sound-idea that might first have inspired it.

> Now here is the problem set before those who endeavour to discover the imitative roots of language. They have to decipher the most weather-worn records of the human race—records subject to such influences as those which brought Tooley Street out of St. Olave Street, Jour out of Dies, and offspring so unlike each other as Bishop and Evêque [*sic*] from the same immediate parent. If we consider the length of time during which these obliterating influences have been at work upon language, we shall be surprised, not at the wide lacunae in the chain of evidence which we extract from out witnesses, but that the faint and hesitating accents in which they necessarily speak can afford us any sound link whatever.[56]

This argument surely operated at two levels, one open and obvious, the other covert. Ostensibly, it defended the imitative theory of language, yet it also set up a tacit parallelism with the idea of biological common descent. Julia Wedgwood thus intertwined two logically distinct issues: the question of the origin of speech and the linguistic analogy. She also presented the first case of an unstated polemic by analogy; further instances of this are cited in chapter 4.

In her concluding paragraph, Wedgwood raised a paean to uniformitarianism, and here again she conveyed a double meaning. Geology, she said, had taught its students to discover keys to the past in observable, ongoing processes, "and to see in every shower of rain a specimen of the forces to which the present state of our globe is owing. The study of language, we doubt not, is destined to achieve an analogous triumph over the weakness of our imagination."[57] In other words, a primary requirement for sound scientific reasoning was the conquest of a weak imagination. This thesis, which was implicit in Charles Lyell's case for uniformitarian geology, Wedgwood made fully explicit in her treatment of language.

At the very time when Julia Wedgwood's review appeared, Max Müller's lectures inspired yet another analogy-oriented discussion. This was an analysis of the linguistic image itself, carried on in the private correspondence between Charles Darwin and Asa Gray. Along, it seems, with the entire educated public on both sides of the Atlantic, Darwin and Gray each read Müller's new volume on language. They then exchanged opinions of the book and, in so doing, produced a very curious discussion. Primed by jousts with his Harvard colleague Agassiz, Gray paid close attention to Müller's stress on community of linguistic descent. He therefore told Darwin that Müller's work contained useful ideas. "Perhaps what has interested me most is, after all, his perfect appreciation and happy use of Natural Selection, and the very complete analogy between diversification of species and diversification of language. I can hardly think of any publication which in England could be more useful to your cause than this volume is or should be." Gray repeated for emphasis, "Depend on it, Max Müller will be of real service to you."[58]

Strangely enough, Darwin failed to recognize in Gray's remarks a description of his own illustrative argument. His reply was therefore surprisingly skeptical. He agreed that Max Müller's book was "extremely interesting" but found "the part about the *first* origin of language much the least satisfactory. It is a marvelous problem. [But] I cannot quite see how it [the book] will forward 'my cause,' as you call it."[59]

Part of this misunderstanding stemmed from Gray's having said that he was impressed by two things, the "very complete analogy" between linguistic and biological diversification, and—what he mentioned first—Müller's apt use of the natural selection principle. Apparently, Darwin overlooked Gray's reference to the analogy and seized upon his mention of natural selection, which, as he clearly knew, had appeared as a preliminary to Müller's mystical explanation of the origin of speech.[60] Müller had argued that humanity's original language must have contained a plethora of synonymous root words, for the oldest known tongues possessed multiple synonyms for basic terms such as *man*, *earth*, and *sun*. Borrowing from Darwin, Müller said that those words gaining the most popular approval, hence usage, had been naturally selected to survive. This "struggle for life" had eliminated the redundant words and had left surviving the major term passed down to each of the modern languages.[61] Darwin, no doubt, noted well this shrewd yet uncongenial use of his selection idea, and this knowledge predisposed him to discount any notion that Müller's lectures might prove useful to his cause.

Thus far, Darwin's reply to Asa Gray was misdirected, yet for understandable reasons. His addition remarks, however, are genuinely perplexing. In passing, Darwin mentioned what he considered an alternate use of philological themes. He did not see how Max Müller's ideas could advance his cause, "but I can see how any one with literary talent (I do not feel up to it) could make great use of the subject in illustration. What pretty metaphors you could make from it!" Evidently, Darwin here regarded the illustrative appeal to language as a novel idea, and this only a few years after using just this kind of metaphor at several points in *The Origin of Species*. It had been an even shorter time since he had reveled in Gray's report concerning Louis Agassiz's debacle at the American Academy meeting. Nevertheless, Darwin turned again to the problem of the first origin of speech, adding, "I wish some one would keep a lot of the most noisy monkeys, half free, and study their means of communication!"[62]

Exasperation showed in Asa Gray's response. Gray chided Darwin for misconstruing his praise of Max Müller's volume, for he had not meant to refer to Müller's views on the origin of speech. "Surely you can't wonder that the attempt to account for the 'first origin of language,' or of anything else, should be the 'least satisfactory' " part of a book. Ironically, Darwin's failure to grasp Gray's point forced Gray to spell out his true meaning, and, in so doing, to present an impeccable description of what Darwin himself had already attempted in *The Origin of Species*. "The use that I fancied could be made of Max Müller's book, or rather of the history of language, is something more than illustration, but only a little more; that is, you may point to analogies of development and diversification of language, of no value at all as evidence in support of your theory, but good and pertinent as rebutting objections urged against it."[63]

Gray was right to say that the findings of historical philology were of no value as evidence in support of Darwin's theory yet that they still could be used for "something more than illustration." He saw that philology's value lay in "rebutting objections," precisely the end to which Darwin used such material in trying to show that his descent theory was not unreasonable. This was the modest analogical function that not only Darwin but also George H. Lewes, T. H. Huxley, and the Wedgwoods had envisaged.

It is mildly puzzling to find Asa Gray forgetting that Darwin had deployed such images for this very purpose; penciled markings in Gray's copy of the *Origin* show that he had indeed taken note of the book's several linguistic analogies.[64] What is thoroughly mystifying, however, is that Darwin should have forgotten his own use of these images.[65] The best explanation is that Darwin was immersed at this moment in the "mystery of mysteries," the origin of humankind. At the time of his exchange of letters with Gray, he was following the news of England's "apes and angels" controversy and was anxiously awaiting new works on man by Huxley and Lyell.[66] Hence, as regards anything linguistic, Darwin's preoccupation with the origin of speech. In mentioning Max Müller's lectures to his friend Hooker, he accordingly repeated what he had told Gray: "the part about [the] first origin of language seems the least satisfactory part."[67]

Asa Gray had made a further point when he wrote to Darwin concerning Müller's lectures, although the full significance of his remarks would not come out until later. There was an additional way, Gray said, in which the history of languages could prove useful to Darwin's cause. "I see also with what great effect you may use it in our occasional discussion about design; *indeed I hardly see how to avoid [a] conclusion adverse to special design*, though I think I see indications of a way out."[68] This was a pivotal comment, for here Gray introduced Darwin to the linguistic image's ambivalent message. Gray suggested that Darwin could use the analogy not only to support his basic theory, that is, to make plausible the idea of transmutation, but that he might also use it to emphasize a distinctly naturalistic interpretation of that theory. Gray, the theistic Darwinist, thus perceived what to him was a threat, for he saw the analogy suggesting a conclusion "adverse to special design." His remarks on this subject would have a telling effect, for they would spark Darwin's awareness of the subtle lights and shadows hidden within the language-species comparison. This awareness, in turn, would affect Darwin's handling of the design issue. Gray did not now elaborate, but he and Darwin would return to this problem in several months' time.

Meanwhile came the event that goaded that future discussion, the appearance of Charles Lyell's *Geological Evidences of the Antiquity of Man*. With an entire chapter devoted to language, this work catapulted the language-species analogy into prominence as never before. It also made the linguistic image the chief vehicle for expressing the philosophical disagreement within Darwin's scientific circle.

Linguistic Natural Theology

Antiquity of Man came as a disappointment to Darwin, and relations between him and Lyell were strained during the period surrounding the book's publication. Darwin had hoped that this work, projected as a treatise on primeval man, would go the whole way in applying his transmutation theory to humans. The arrival of an advance copy showed otherwise.[69] Lyell's *Antiquity* made a case, chiefly, for humanity's long prehistory, far longer than suggested by the traditional Mosaic chronology. Lyell turned, however, in the book's final chapters, to the topic of Darwin's descent theory.

As a bridge, he began by discussing race. He acknowledged the persistence of distinct racial types, confirmed by the human figures depicted in four-thousand-year-old paintings on the walls of Egyptian temples. Yet he made a case for the gradual modification of those types from a common origin, as allowed by the lengthened human time frame. As the historian Edward Lurie has noted, Lyell produced in this context a penetrating exposé of Louis Agassiz's ethnological polygenism.[70] That is, he noted the deeper significance of the fact that "some zoologists of eminence" held the several human racial types to be "primordial creations." "Were we to admit, say they, a unity of origin of such strongly marked varieties as the Negro and the European, . . . that, in the course of time, they have all diverged from one common stock, how shall we resist the arguments of the transmutationist, who contends that all closely allied species of animals and plants have in like manner sprung from a common parentage?"[71]

It was not, of course, a matter of human monogenesis actually demanding a transmutationist theory of all plant and animal species. Rather, what Agassiz feared was the analogic impression that might be made on uncommitted minds. Repeating what he had written in his journal, Lyell observed that Agassiz's commitment to polygenism possessed "at least the merit of being consistent with itself, and relieves the opponents of transmutation from the dilemma of explaining why, if so great a divergence from a parent type as that of the white man and Negro can take place, a like modifiability should not be able, in the course of ages, to go a step farther, and give rise to differences of specific value [i.e., of species]."[72] By thus analyzing Agassiz's complaint, Lyell placed the race-species parallelism on favorable display.

Antiquity supported Darwin's theory in other ways as well. Not surprisingly, Lyell especially backed Darwin's explanation of the gaps in the fossil record.[73] Yet on the whole, his endorsement was tepid. Particularly disappointing to Darwin, Lyell's final chapter (chap. 24) stopped short of applying the descent theory to humanity. And on the book's final page, Lyell commended Asa Gray's theistic Darwinism and suggested that a Prime Mover was necessary to begin the evolutionary process.

Sandwiched between these passages for and against Darwin's own viewpoint, and supplying a transition from the one to the other, there appeared the book's penultimate chapter, "Origin and Development of Languages and Species Compared." Here one finds perhaps the high-water mark of scientific discourse by analogy, for this chapter held the heart of Lyell's argument and was actually the closest he ever came to clarifying his views on Darwinism. Appropriately, the linguistic images here would be more complex than anything that had yet appeared in print. An elaboration of Lyell's journal entries from several years earlier, chapter 23 of *Antiquity* carried two messages. In part, Lyell used linguistic analogies in the same spirit that Darwin had in the *Origin:* he tried to entice philologically informed readers—and by then there were many, thanks to Max Müller—to suspend their disbelief.[74] Yet Lyell's analogizing also served the contrasting function of attempting to convince Darwinians that a profound mystery lurked behind natural selection.[75] In each case, to use Martin Rudwick's phrase, Lyell bid to "convert the scientific imagination" of his readers.[76] He set out first to win the skeptics.

Lyell began by acknowledging his debt to Max Müller, the author of "the most improved version" of the theory positing an ancestral Indo-European tongue.[77] Yet he added a new wrinkle: although it was clear that these languages had gradually diverged from a single source, it was also clear that they had done so much more quickly than had mankind's long-distinct racial types. Implicitly, then, Lyell rejected the superimposed linguistic and racial genealogies in the ethnological illustration in chapter 13 of the *Origin*. The gist of Lyell's argument was a new three-level comparison involving language, race, and biological species. While races diversified at a much slower rate than did cognate languages, a new species would take "an incomparably longer period" to evolve than would a new race. Hence if languages and species were comparable in their developmental arrangement, they were not so in time scale. This brought Lyell to a striking conclusion: "A philologist, therefore, who is contending that all living languages are derivative and not primordial, has a great advantage over a naturalist who is endeavoring to inculcate a similar theory with regard to species."[78]

With the philologist's advantage in mind, Lyell conjured up a thought exercise. It would not be "uninstructive," he said,

> in order fairly to appreciate the vast difficulty of the task of those who advocate transmutation in natural history, to consider how hard it would be even for a philologist to succeed, if he should try to convince an assemblage of intelligent but illiterate persons that the language spoken by them, and all those talked by contemporary nations, were not modern inventions, moreover that these same forms of speech were still constantly undergoing change, and none of them destined to last for ever.[79]

In other words, although the philologist normally held an advantage over the naturalist in demonstrating descent with modification, it would be hard, under these hypothetical conditions, "even for a philologist" to make his case. This was so because an illiterate community would doubtless meet the idea of linguistic mutation with certain reasonable objections: Did not spoken language appear to hold relatively stable within each individual's lifetime? What evidence was there of past variation? And, most important, where were the traces of intermediate dialects, both synchronic and diachronic? "How comes it," they would ask, "that the tongues now spoken do not pass by insensible

gradations the one into the other, and into the dead languages of dates immediately antecedent?"[80]

Lyell's response showed that, although it might be difficult for a philologist
to sustain his case, it would not be impossible. Indeed, philologists had already
established such certainties. First, they had found that none of the modern
European languages was more than one thousand years old: modern English
speakers, for instance, could not readily understand Anglo-Saxon. Second,
scholars had proved that most modern languages had emerged from tongues
formerly spoken in roughly the same geographic regions where their present-
day speakers reside. Finally, they had noted that the ancient languages had
"passed through many a transitional dialect before they settled into the forms
now in use." In this context, Lyell seconded Darwin's claim that the line of
demarcation between a language and a dialect, like that between a species and a
breed, was nearly impossible to establish. Deferring to his hypothetical ques-
tioners, he acknowledged that "many and wide gaps" appeared between the
remnants of living and dead languages. Yet he gave reasons for this circum-
stance, noting the poor preservation, or lack, of old texts showing the stages of
a given language's development—obvious references to the "imperfection of
the geological record."[81]

Again speaking for the illiterate community, Lyell wondered whether the
"trifling" changes each generation saw in its own speech could possibly lead to
the tremendous number and diversity of languages throughout the world. His
reply was simple: any extent of change was possible given sufficient time. By
the standard of the linguistic analogy, then, time was on Darwin's side, for that
analogy was now wedded to the lengthened human chronology. Lyell thus met
the most common objections to the idea of biological mutability by explaining
how similar obstacles had been overcome in the investigation of language.[82]

Apart from the new emphasis on abundant time, the linguistic themes seen
so far in Lyell's chapter had appeared already, albeit with less elaboration, in *The
Origin of Species.*[83] Yet Lyell soon left these familiar arguments behind and sailed
into uncharted waters. Quietly and without warning, he reversed the direction
of the analogy: instead of using language to represent biological phenomena,
he now used Darwinian categories to explain the history of language. Thus was
introduced an entirely new phase of Lyell's argument, although the purpose of
the change was not immediately clear.

Lyell began the new tack by considering the familiar problem of lexical
fecundity, a theme probably borrowed from Max Müller's *Lectures on the Science
of Language* (1862). Lyell observed that, although most new words, idioms, and
phrases introduced into a given language were of only "ephemeral duration,"

their total number might within a century or two compare with that language's entire long-term vocabulary. What prevented such superabundance in actual usage? What laws, Lyell asked, might govern "not only the invention, but also the 'selection,' of some of these words or idioms, giving them currency in preference to others? For, as the powers of the human memory are limited, a check must be found to the endless increase and multiplication of terms." Here was the Malthusian principle applied to language, human memory functioning in the same way that limited food supply checked a population's "endless increase and multiplication." That this phase of Lyell's discussion constituted a reversal of the original analogy was evident from its application of this and other Darwinian concepts to the explanation of linguistic change. Lyell spoke of

> fixed laws of action, by which, in the general struggle for existence, some terms and dialects gain the victory over others. The slightest advantage attached to some new mode of pronouncing or spelling, from considerations of brevity or euphony, may turn the scale, or more powerful causes of selection may decide which of two or more rivals shall triumph and which succumb. . . . Between these [local] dialects, which may be regarded as so many "incipient languages," the competition is always keenest when they are most nearly allied, and the extinction of any one of them destroys some of the links by which a dominant tongue may have been previously connected with some other widely distinct one. It is by the perpetual loss of such intermediate forms of speech that the great dissimilarity of the languages which survive is brought about. Thus, if Dutch should become a dead language, English and German would be separated by a wider gap.[84]

There followed a long series of comparisons in this vein, all designed to show that "the doctrine of gradual transmutation" was "applicable to languages." A few examples will suffice. Just as a transmutationist would predict, (1) each living tongue was preceded by a "closely allied prototype," internal evidence of which could be found in languages now spoken. (2) The predecessors of contemporaneous related languages converged back to a single geographic point of origin; surely no language could have two birthplaces.

Then came two especially telling comparisons. Like biological organisms, (3) languages manifest continuity over time, displaying "the tendency of each generation to adopt without change the vocabulary of its predecessor." Lyell compared this continuity with the force of inheritance in biology. Yet languages also exerted (4) "inventive" force: the human tendency to coin new words and to readapt old ones to changing conditions "answers to the variety-making power in the animate creation."[85] As everyone knew, the ability of organisms to insure inheritance and to produce variations was a mystery that Darwin's theory had done little to solve. Similarly, while the Darwinian schema applied to the growth of language in many ways, it left much unexplained. Hence, Darwinian categories were indeed applicable to languages, not only in the positive insights they provided but in their explanatory limitations as well.

It is important to see that the foregoing discussion was hardly demanded by the logic underlying the initial part of Lyell's chapter. At first, Lyell had used linguistic themes to suggest that descent with modification was a real possibility. When he reversed direction, however, suggesting the application of Darwinian categories to linguistic development, he did this ultimately to raise doubts about any purely naturalistic rendering of Darwin's theory. Even though disenchanted nature might explain much, Lyell hinted, it could not explain all—especially variation and inheritance. In other words, Lyell's "Darwinian" linguistics pointed up the need for something more.

Pursuing this idea further, Lyell produced remarks modeled on a passage in his journal ("When we have got rid of the Garden of Eden. . . . "). All of the mechanisms governing the survival of certain words and dialects may be well understood, he argued, yet historical philology was still plagued by unanswered questions: "When we have discovered the principal causes of selection, which have guided the adoption or rejection of rival names for the same things and ideas, rival modes of pronouncing the same words and provincial dialects competing with one another—we are still very far from comprehending all the laws which have governed the formation of each language."[86] That is, there were certain laws of language formation that could not be accounted for by a strictly Darwinian view.

Proceeding to the next stage in his argument, Lyell left behind the rivalry among competing words and dialects and turned his attention to a more enigmatic aspect of language development: the complex grammatical systems found in nearly all of the world's tongues. These systems of "rules and inflections" had been slowly constructed through the infinitesimal contributions of each speaker. "The savage and the sage, the peasant and the man of letters, the child and the philosopher, have worked together, in the course of many generations, to build up a fabric which has been truly described as a wonderful instrument of thought." Accordingly, language was like an elaborate "machine, the several parts of which are so well adjusted to each other as to resemble the product of one period and of a single mind."[87] The idea of theistic design here is obvious enough. Yet there was more to be added before Lyell's full case could be made: the slow and piecemeal construction of grammatical apparatus involved "a profound mystery, and one of which the separate builders have been almost as unconscious as are the bees in a hive of the architectural skill and mathematical knowledge which is displayed in the construction of the honeycomb."[88]

These remarks about unconscious constructive action came near the end of *Antiquity's* language chapter. Before examining the goal to which they were directed, it is useful to look at the probable influences behind Lyell's thinking at this point. His sources were eclectic, for he wove together threads from the two main schools of natural theology, the functional-adaptive and the transcendentalist. The former took precedence, for language was a useful "instrument of thought." Yet Lyell's remarks also suggested an idealist overlay: grammar's ordered patterns bore the impress not of the speakers themselves but of some superintending intelligence.

Lyell would have found this hybrid design thesis in a less celebrated strain of British natural theology, one not actually inspired by nature but by human society. Indeed, similarities of concept and detail suggest that he borrowed directly from Richard Whately's *Introductory Lectures on Political Economy* (1831)—which had been reprinted in 1855. Whately, who was the Anglican bishop of Dublin, noted that beneficial results were often brought about "by the joint agency of persons who never think of them, nor have any idea of acting in concert." Those who navigate ships, for instance, took no more thought for national wealth and commerce "than the bee has of the process of constructing a honeycomb." In his most detailed illustration, Whately described the daily economy of the city of London, a vast system of exchange and sustenance in which thousands of individuals unknowingly cooperated.

In the spirit of Adam Smith's "invisible hand," Bishop Whately's political economy revealed rational agents acting out of self-interest yet combining "as regularly and as effectually [for] the accomplishment of an object they never contemplated, as if they were merely the passive wheels of a machine." Even the most skilled human administrator could not have created such a system; here, rather, one found "the same marks of contrivance and design, with a view to a beneficial end, as we are accustomed to admire (when our attention is drawn to them by the study of Natural Theology) in the anatomical structure of the body, and in the instincts of the brute creation."[89]

Nevertheless, even here one finds brush strokes from what Baden Powell described as nature's "vast scheme of universal order and harmony." In his "Spirit of the Inductive Philosophy" (1855), Powell pointed to the geometrically precise cells of honeybees as evidence of nature's mental blueprint, "independent of the idea of mere utility."[90] Whately just as plausibly used the beehive as an analog of unplanned social cooperation, a manifestation of universal mind and utility both. He thus did something new: by making his subject human society rather than, say, the human eyeball, he was able to bring together the two opposing schools of natural theology. In short, natural theology in the social realm allowed idealism and teleology to be combined.

Charles Lyell needed only to transfer Whately's socioeconomic themes to the realm of language. As we saw above, Lyell too adopted the beehive illustration, using it as a secondary analogy to highlight the unplanned symmetry of grammatical systems. (Lyell probably alluded to the beehive for a more direct reason as well, knowing that Darwin had found this phenomenon a stumbling block to his theory of blind and accidental natural selection.)[91] Yet again, grammar could also suggest purposive design, for it served the ends of communication. To summarize, language presented an instance of human behavior serving a utilitarian purpose, yet its strict phonetic laws and grammatical orderliness, phenomena beyond the purposeful intentions of its speakers, suggested an added canopy of mental design.[92]

Having built these themes into his *Antiquity* chapter, Lyell then finished out his argument and made plain the thesis that had capped off the discussion of language in his scientific journal of 1860. If the best modern philology revealed such ultimately mysterious historico-grammatical patterns, he asked, what then should we expect from "our attempts to account for the origin of species," in which

> we find ourselves still sooner brought face to face with the working of a law of development of so high an order as to stand nearly in the same relation as the Deity himself to man's finite understanding, a law capable of adding new and powerful causes, such as the moral and intellectual faculties of the human race, to a system of nature which had gone on for millions of years without the intervention of any analogous cause. If we confound "Variation" or "Natural Selection" with such creational laws, we deify secondary causes or immeasurably exaggerate their influence.[93]

Here, at last, Lyell showed his hand: although couching his idea in terms of nature's laws, he made public his belief in the continuous introduction of "new and powerful causes," as well as an increasing share of rationality, in the development of both languages and living organisms. In languages, this occurred through increased lexical refinement; in organisms, it corresponded to an increasing complexity over the course of biological history, the apex appearing in human rationality and moral sense.[94] Yet the format here was different from that in Lyell's journal. In the *Antiquity* chapter, Lyell added an explanation of linguistic development in Darwinian terms as far as he thought possible. In this way he was able to suggest the limits of these mechanisms. By the time readers arrived at the part conveying Lyell's theistic idealism, they had been shown, presumably, that something more than naturalistic forces were needed.

Charles Lyell thus set forth his long linguistic response to Darwin, far outdoing *The Origin of Species* in the use of such tropes. At the beginning, Lyell used these in the same way as Darwin did: stressing the achievements of modern philology in spite of its dependence upon scanty evidence, he presented convincing reasons not to reject Darwin's hypothesis. Yet in the end he turned the tables on Darwin by pointing out the limitations of even so successful a scholarly venture as philology.

A Broken Circle

Darwin wrote to Asa Gray and J. D. Hooker early in 1863, complaining bitterly about the "excessive caution" of Lyell's *Antiquity*. He predicted that the book would "not serve as guide to anyone." There were, however, a few positive features, for Lyell had convincingly outlined some of the more important aspects of Darwin's theory. Surprisingly, Darwin included among these strong points the new book's chapter on language. After airing his complaints to Gray, Darwin added, "Lyell was pleased, when I told him lately that you thought that language might be used as [an] excellent illustration of [the] derivation of species; you will see that he has [an] *admirable* chapter on this." To Hooker, Darwin described Lyell's chapter as "most ingenious and interesting." Then Darwin wrote to Lyell himself, frankly expressing his disappointment with the book as a whole. And yet, he said, he admired the skill with which his friend had "selected the striking points and explained them. No praise can be too strong, in my opinion, on [sic] that inimitable chapter on language in comparison with species."[95]

Darwin was fully aware of that chapter's message that divine intelligence had superintended the evolutionary process. Yet he chose to overlook this and to focus on the "admirable" qualities. Indeed, this appreciation may not have been hard to muster. After all, much of Lyell's chapter ratified the plausibility-enhancing analogies that Darwin had used in the *Origin*. And the overtly theistic passage at the chapter's conclusion was brief, less than a page in length.

After this, Lyell returned to the fold: he advised readers not to undervalue the importance of Darwin's ideas. "All our advances in the knowledge of Nature," he said, had consisted of adopting similarly controversial new hypotheses, "and we must not be discouraged because greater mysteries remain behind wholly inscrutable to us."[96] In a resounding finale, Lyell used a philological image to commend uniformitarian methodology, not only in geology but also in biological investigations. In this light, Darwin's enthusiasm for Lyell's language chapter makes sense. Its prodesign subtext notwithstanding, that chapter had gone further toward suggesting an outright acceptance of transmutationism than had anything else in Lyell's new book. Darwin's disappointment was reserved mainly for his friend's doubtful response on the subject of human descent.

After Darwin vented his frustration on the latter score, Asa Gray remonstrated on Lyell's behalf. He told Darwin that it was unreasonable to complain about *Antiquity* not backing his theory more forcefully, especially since its guarded approach was likely to have "much success in disarming prejudice. And this is all you could ask."[97] Gray declared Lyell's presentation "first-rate," adding, "It is just about what I expected, and is characteristic of the man." Gray then gave his opinion of the chapter on language. His allusive remarks on this subject will require close analysis, as will Darwin's apparently contorted reading of those remarks. This brief pair of comments, buried within the substantial Darwin-Gray correspondence, marked a significant change in Darwin's understanding of the linguistic analogy.

After defending *Antiquity* as a whole, Gray added this cryptic observation: "The chapter on language makes the points I supposed would be made, or some of them, but only dips in, leaving more to be said. But this is rather ticklish ground, for, if we are not careful here, you would get the better of us in this field *quoad* design."[98] What did Gray mean by this? First of all, he suggested that Lyell could have done more with the analogy; how, he did not say. He hinted, however, that this lack on Lyell's part was really for the best. For, just as he had done six months earlier, in discussing Max Müller's *Lectures on the Science of Language,* Gray here suggested that the language-species analogy presented a latent threat: "But . . . design." In other words, it was just as well that Lyell had pushed the linguistic analogy no further lest it raise issues harmful to a design interpretation of Darwinism. Again, Gray did not explain why he thought this was so. What is clear, however, is that he drew a line between Darwin, on the one hand ("you"), and himself and Lyell on the other ("we" and "us"), signaling his basic sympathy with the British geologist. Even so, by emphasizing the (unspecified) danger of pressing the analogy further, Gray effectively discounted Lyell's efforts to so convey a prodesign view of evolution. Indeed, Gray suggested that Lyell's analogizing actually tended to undermine a design argument.

But how exactly did Lyell tend to do this? And in what sense did he "leave more to be said"? Darwin himself answered these questions, supplying a plausible reading of Gray's lamentation. Darwin's thoughts were not fully revealed in his reply to Gray, to whom he simply declared: "How clever and original and candid your remark about Language and Design." At the same time, however, Darwin sent Gray's letter on to Hooker, and on this occasion he added an astonishing interpretation of what Gray had said. "That is a clever remark in Gray's letter about [the] origin of language telling against each trifling variation being designed; Lyell shirked this point, which I urged him to grapple with."[99]

The striking thing here is Darwin's claim that Gray had spoken of the "origin of language telling against each trifling variation being designed." Strictly speaking, Gray's remarks on Lyell had suggested nothing of the kind, for they mentioned neither linguistic origins nor small variations. Yet this was not just another misreading on Darwin's part of a letter from Asa Gray. It is true that neither Lyell's *Antiquity* chapter nor Gray's comments on it had addressed the actual origin of language. The seeming incongruity on this point stemmed from Darwin's use of the word *origin* to include what should be called "development." To Darwin, of course, origin *was* development, the obvious case being the origin of species. When Darwin wanted to restrict the term, he made this plain: for example, he had twice expressed dissatisfaction with Max Müller's theory explaining the "first origin of language," that is, how the earliest humans had for the first time produced articulate speech.[100] Now, however, he omitted the qualifier *first,* and used *origin* to indicate the ongoing emergence of new languages from tongues formerly spoken.

The real mystery lies in the second part of the statement, where Darwin said that the manner in which languages evolved precluded "each trifling variation being designed." How did he find this message in Gray's remarks? To repeat, Gray had said of Lyell's analogizing: "this is rather ticklish ground, for, if we are not careful here, you would get the better of us in this field *quoad* design." Fortunately, Gray's comment did not appear in a vacuum but recalled one of the things Gray himself had said six months earlier concerning Max Müller's *Science of Language.* Gray had frankly suggested that comparative philology could prove useful to Darwin's position—though not his own—"in our occasional discussion [i.e., debate] about design; indeed I hardly see how to avoid [a] conclusion adverse to special design."[101] This remark was most likely inspired by Müller's application of natural selection, or as Müller called it, *natural elimination,* to the early development of language.

Asa Gray would have noticed the appearance of nearly the same idea in Lyell's *Antiquity* chapter, concerning how most of the words, idioms, and phrases ever invented had been eliminated in the lexical struggle for existence. In both languages and organisms, most variations would never be selected for survival; most, rather, would be left behind on the threshing floor of evolution.

Neither Müller nor Lyell had been bothered by this idea: Lyell had alluded to it only as a preliminary to his point about the "creational laws" superintending the course of linguistic change.

Yet in his eagerness to stress the regularity and order of language's finished structure, Lyell had overlooked the colossal waste of its raw materials; he had not faced up to the way most linguistic innovations were shredded away wholesale like the random variations that fed the maw of Darwinian selection. Hence Gray's remark that Lyell "only dips in, leaving more to be said." For his part, Gray had addressed the issue of variation. Even more skeptical than Lyell was of the idea of blind selection, Gray had argued in his *Atlantic Monthly* articles that variation itself had been "led along certain beneficial lines." If this were not the case, and a host of variations were sacrificed as in the case of language, then the language-species analogy actually undermined any theistic argument. Thus Gray saw that this aspect of the parallelism was "adverse to special design."[102]

These are the things, I argue, that Darwin must have read in Asa Gray's terse evaluation of Lyell's *Antiquity* chapter. Darwin had understood Gray as candidly admitting that, if one stressed the idea of "trifling variations," most of which were wasted, it would undercut any argument for the harmonious or beneficent character of transmutation. The result was Darwin's conclusion that what was known of early lexical development told "against each trifling variation being designed."

We now can see how Lyell's and Darwin's opposed interpretations of natural selection reflected their pick of social philosophers. Even though cast in linguistic terms, Lyell's *Antiquity* argument sought to recapitulate Richard Whatley's response to Thomas Malthus. Lyell had meant to write in the optimistic spirit of Whatley's harmonious social ecology, with its city and market serving as beneficent distributive mechanisms. This viewpoint followed in the tradition of Adam Smith and the equitable dealings of the invisible hand, not only a matter of economic theory but also a justification of God's ways to man. Countering Malthus's picture of disharmony between humans and the natural environment and the impediments to progress thus created, Whatley touted the benevolent force of unintended social cooperation.[103] Even so, the spirit of Malthus could not help but also enter in whenever someone employed the idea of natural selection—as Lyell did extensively in the second half of his language chapter. Hence Darwin's ties to the pessimistic parson inevitably crept into Lyell's analogizing. Lyell had wanted to highlight the explanatory limitations of natural selection. He had intended to show that, despite the mundane workings of this mechanism, supernatural forces were required to propel the overall development of both language and nature; and, in addition, that both of these realms manifested increasing rationality over time. Yet in conveying this through the analogy with language, he let the genie out of the bottle: he inadvertently contaminated his theistic vision with the notion of competitive destruction—his argument thereby producing an unintentionally baleful subtext.

In pointing out this weakness in Lyell's argument, Asa Gray betrayed no loss of confidence in his own views about design in nature. Still, his aid and comfort to Darwin's side of the analogic tug of war was substantial. And in a sense, more than an analogy was involved. To Darwin's thinking, the language analogy now could be seen as addressing the vexing charge that many critics (Gray included) had been making that natural selection was not a creative force and that variation could be such only if it supplied evolution with a positive direction. The natural development of language, however, presented a case in which selection acted upon variations that were themselves apparently as random and nonteleological as Darwin ever could have desired. Ironically, then, Lyell's kindred spirit Asa Gray helped Darwin discover a linguistic message the very opposite of the one Lyell had intended. Not only did Gray hint that Lyell's prodesign version of the analogy was ineffective, but he also appears finally to have convinced Darwin that, for this very reason, the analogy could prove more useful to his cause than Darwin himself had at first realized.

Did others outside Darwin's circle perceive these issues in *Antiquity*'s twenty-third chapter? Measured by published reactions to the book, most of them written by geologists, Lyell's discussion of language exerted small influence. Some reviews, like the one in London's *Athenaeum,* passed over the section on transmutation entirely; others just skipped the language chapter. And those which did mention that chapter were unimpressed. As the *Edinburgh Review* declared, "The argument from the analogy between the time required to introduce a new word into a language, and a new species into the chain of being, is rather rhetorical than apposite, and is not, we believe, even new." The Tory *Quarterly Review* merely noted that "the controversies regarding the origin and varieties of language . . . offer so much analogy to the questions concerning organic life." And in New England the geologist C. H. Hitchcock rather obtusely declared that the linguistic comparison "very illogically derives an argument for the origin of species by selection from the origin of dialects by means of human selection, thus confounding together mental and physical laws."[104] Other reviewers on both sides of the ocean similarly discounted the analogy as "too shadowy, figurative, and far-fetched, to have any logical force whatever," apparently oblivious to the way it also qualified Darwin's theory. Hence if respondents to Lyell's *Antiquity* said anything about its language chapter, they tended to concentrate on its protransmutationist phase, and then only to discount its polemical force.[105]

One of Lyell's more prominent readers, however, was more appreciative of the power of analogy. Now near his seventieth year, Lyell's old philosophical combatant William Whewell wrote a complimentary letter, praising *Antiquity's* language chapter and referring to his own bringing together of the palaetiological sciences years before in his *History of the Inductive Sciences* (1837) and other works. Whewell cautioned, however, that although Lyell's analogies admirably illustrated the "difficulties and the solutions" with respect to Darwin's descent theory, it did not amount to an argument for either side. This challenge precipitated a minor debate.

Whewell reiterated his long-standing opinion that smooth uniformitarian change, such as Lyell's creational laws produced, could never account for the beginnings of things, languages included, and that "catastrophic" influences must be supposed. Lyell reminded Whewell that all existing languages had gradually developed from those formerly spoken, without any forces of change superior to those found at present. Whewell replied that, even with the lengthened human chronology, explosive beginnings would have been needed. English, he said, had sprung up quickly, a little less than a century after the Saxon conquest, and had undergone "little essential change" (a preposterous claim) since that time.[106] Whewell's argument was in its own way as quixotic on the linguistic side as were Louis Agassiz's antitransmutationist philological beliefs.

Despite these various negative responses, one should not conclude that the *Antiquity* chapter failed to exert significant influence. As it turned out, its most important and attentive audience was the one Lyell had probably aimed at all along—Darwin himself and his immediate circle, including the Young Turk T. H. Huxley. These were the ones who especially needed to be convinced that natural selection could not fully explain the development of biological phenomena. Yet the effect was the opposite of what Lyell had intended. Because of Asa Gray's greater consistency of thought on the problem of variation, the weakness of Lyell's analogy became apparent to Lyell's most important reader. The ironic result was to reconfirm to Darwin the aptness of that analogy for recommending his own viewpoint. It would lead him toward redeploying such material a number of years later, in a peculiar and digressive passage in *The Descent of Man* (1871).

A Truncated Image

We return, finally, to the issue with which we began, the relationship between language and race. As was noted at the beginning of this chapter, the 1850s produced strong arguments for the logical separation of these two phenomena.

Max Müller campaigned for this early on, and T. H. Huxley made the case from the zoologists' side a decade later. Reviewing the state of ethnology in 1865, Huxley acknowledged the usefulness of linguistic research "as an adjunct," yet he denied it the leading role in determining community of racial descent. Close similarity of languages did suggest a presumption in favor of ethnological unity among the peoples speaking those languages, yet this could never actually prove such a relationship and in some cases could mislead. The Fijian people, for example, were physically allied to the "Negritos" of New Caledonia, even though their language was Polynesian. "But if languages may be thus transferred from one stock to another, without any corresponding intermixture of blood, what ethnological value has philology?" Huxley asked. After these considerations appeared in print, the naturalist Huxley and the philologist Müller forged an alliance. As Huxley said, if each pulled from his own direction, "I have hopes we shall be able to get Ethnology and Phonology apart."[107]

Thus ran the logical case against the identification of race and language. The revolution in human chronology loosened that identification even further, for with prehistory immensely lengthened, the question of human unity or diversity was rendered practically unanswerable on the basis of comparative-linguistic evidence. Now the family tree of nations could be regarded as identical with that of languages only in limited spheres, as, arguably, in the case of the Indo-European peoples and tongues. But historical linguistics could no longer hope to prove the unity of humanity as a whole.[108]

Still, as the historian George Stocking points out, even as the time revolution undercut the historico-linguistic argument for monogenesis, it actually helped the anatomical argument. Charles Lyell suggested as much in *Antiquity* in his refutation of Louis Agassiz's polygenism.

> So long as physiologists continued to believe that man had not existed on the earth above six thousand years, they might, with good reason, withhold their assent from the doctrine of the unity of the origin of so many distinct races; but the difficulty becomes less and less, exactly in proportion as we enlarge our ideas of the lapse of time during which different communities may have spread slowly, and become isolated, each exposed for ages to a peculiar set of conditions.[109]

In short, the longer the time available, the more the possibility of gradual divergence having taken place prior to the races' long histories of complete distinctness.

Especially important for our purpose is the way these changes in ethnological perspective affected the analogic representation of species. Because language and racial biology were now uncoupled in fact, they tended to be so figuratively as well. Each factor still held good as analogic representatives of organic species, but only separately; they could no longer serve as intertwined representatives, as they did in chapter 13 of the *Origin*. Indeed, at least one British writer pointedly rejected Darwin's illustration on these grounds.[110]

Yet because monogenesis could still be argued on anatomical grounds, racial descent could still provide a cogent parallel for biological descent. As the pro-Darwinian *Natural History Review* declared in 1864, "All adherents to the modern doctrine of the 'origin of species' must . . . *on its general principles* be convinced of the common origin of the varieties of man."[111] Yet there was now no mention of language as confirming the common origin of man. The effect was to cut the linguistic image loose from any remaining ties to racial or ethnological factors and to stand it on its own feet, thus emphasizing the unmediated parallel between languages and biological species.[112] By the very placement of its discussions of language and race, Darwin's *Descent of Man* would reflect both the real and the analogic uncoupling.

The Convoluted Path to
The Descent of Man

CHARLES LYELL'S *Geological Evidences of the Antiquity of Man* (1863) rendered the language-species analogy both imaginative and conspicuous, certainly more so than any other published work until that time. And to a significant extent, Lyell's discussion was pro-Darwinian. Darwin himself clearly approved, as his letters to intimates such as Joseph Hooker show. But did Darwin regard Lyell's chapter on language an unmixed good? Part of that chapter was not at all friendly toward his naturalistic view of selection. On the other hand, as Asa Gray pointed out, Lyell's analogizing had not successfully vindicated supernatural design either. Yet which of these readings of the linguistic image would prevail in print, in the writings of those who continued to invoke it?

To a large extent the career of the language-species comparison in the 1860s mirrored the rejection of theistic design as an officially sanctioned Darwinian viewpoint. In addition, this period saw protransmutationist analogizing reach new heights of boldness, creating a significant departure from the modest function that Darwin, Asa Gray, and others had envisaged. Yet there also arose a sophisticated new version of the anti-Darwinian linguistic analogy: like Louis Agassiz, only with greater subtlety, a leading Victorian scholar used this image to question the notion of organic diversity emerging from a small number of progenitors. Each of these widely varied forms of the analogy made a number of appearances during the post-*Origin* decade, and a review of the roles they played makes for a convoluted story, filled with ironies and strange alliances. Even so, the climax to which all of this led is clear-cut: the analogy making of the 1860s inspired a virtuoso response in Darwin's second most famous book. This chapter's ultimate goal is to account for that superfluous passage in *The Descent of Man* (1871).

August Schleicher's "Darwinian" Linguistics

A whole new phase of the discourse on Darwinism and language began in the same year that Lyell's *Antiquity* appeared. Its beginning was an unlikely one, coming in the form of an open letter written from one professor to another at a provincial German university. The first German-language translation of *The Origin of Species* appeared in 1860, the work of the paleontologist Heinrich Bronn (1800–62).[1] That event marked a turning point in the career of Ernst Haeckel (1834–1920), a young associate professor of zoology at Jena's Friedrich Schiller Universität. Haeckel would go on to become the most outspoken crusader for Darwinism on the European continent, a figure who pushed *Darwinismus* into regions of biological and philosophical theorizing that Darwin himself avoided. Yet Haeckel took perhaps his most momentous step early on, when he presented a copy of the *Origin* (Bronn's 2d ed., 1863) to his older colleague August Schleicher. A leader in the field of comparative philology, Schleicher was also an enthusiastic amateur botanist, and so he seemed a likely candidate for the spell of Darwinism. Duly impressed with Darwin's book, he communicated his reaction to Haeckel in a published letter, *Die Darwinsche Theorie und die Sprachwissenschaft* (The Darwinian theory and linguistic science, 1863).[2]

Addressing the young scientist, Schleicher described the surprising effect of what he had read. "In supposing that Darwin's 'Origin of Species' would please me, you were thinking no doubt, in the first place, of my amateur gardening and botanizing. . . . Yet, my dear friend, you were not altogether on the right track," for "Darwin's views and theory struck me in a much higher degree, when I applied them to the science of language." This readiness to see Darwinian patterns in linguistic phenomena arose from Schleicher's firsthand research. There is no evidence that he had read, or ever did read, Lyell's *Antiquity*. Indeed, there was no need for such external influence on Schleicher's thinking. At least as early as 1860, "that is to say, contemporaneously with the publication of the German Darwin," he had already set forth an evolutionary picture of linguistic change. In other words, Darwin's vision of nature ratified the prior tendency of Schleicher's linguistics. Schleicher therefore declared to Haeckel, "Can you wonder now that the book has made so strong an impression on me?"[3]

The *Origin* inspired Schleicher with the idea that linguistic science (*Sprachwissenschaft*) and biology were entering upon a kind of convergence. Such at least was Schleicher's hope, and that hope was shared to some degree by many nineteenth-century language scholars. Although a young field, the "new philology" was maturing, and its representatives wanted it to assume what they saw as its rightful place among the sciences. Schleicher wanted philologists and naturalists to develop a mutually beneficial interest in each other's work; the results, he hoped, would include both an improved linguistic method and an admission on the part of the naturalists that philology deserved full scientific status.[4] The first naturalist to receive this message was Ernst Haeckel.

The heart of Schleicher's letter was an interplay between two theses, the Darwinian process as applied to language and, conversely, historical linguistics as confirming Darwinism. It was the same dual proposition that had structured Charles Lyell's *Antiquity* chapter, only set forth in plainer terms. First came the "application" thesis: "The rules now, which Darwin lays down with regard to the species of animals and plants, are equally applicable to the organisms of languages, that is to say, as far as the main features are concerned."[5] What botanists and zoologists called the species of a genus were to the philologist "the daughters of one stock-language, whence they proceeded by gradual variation." To illustrate, Schleicher included at the end of his pamphlet a genealogical diagram of the Indo-European languages (see fig. 4.1).

In addition to this descent pattern, Darwin's "struggle for existence" applied as well. Unlike Lyell, Schleicher saw this struggle as taking place not only among individual words but also among entire languages and language groups. Strictly speaking, these larger applications constituted an analogic misuse of Darwin's selection concept, which functioned only among the individuals of a given population. Still, linguistic history manifested a process similar to what really did occur among entire species as well as among human ethnic groups: "Now we observe during historical periods how species and genera of speech disappear, and how others extend themselves at the expense of the dead." North America, for instance, had seen the decay of its native languages and simultaneous spread of the European tongues. After quoting from *The Origin of Species* to show the biological analogs of these and other points, Schleicher concluded that "not a word of Darwin's need be changed here if we wish to apply this reasoning to languages."[6]

Then came the other side of the analogy. Schleicher noted that his Indo-

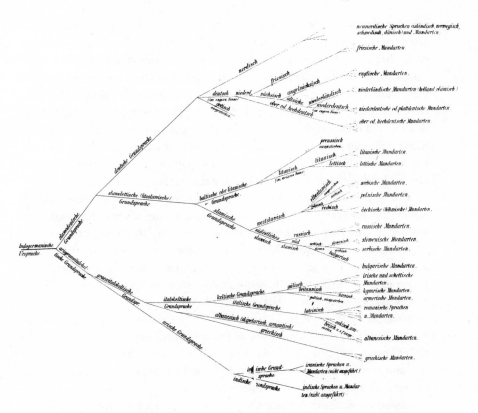

Fig. 4.1. A. Schleicher, genealogy of Indo-European languages from *Die Darwinsche Theorie und die Sprachwissenschaft* (1863).

European linguistic tree closely resembled the *Origin*'s chart depicting the branching descent of species. Yet there was a crucial difference between the two: "One should not forget that the author of the 'Origin of Species' had to draw up an ideal scheme, whereas we [philologists] have represented the actual process of development of a given family."[7] Schleicher suspected that his readers might wonder at the philologists' hubris in reconstructing these linguistic lineages, especially with respect to their earliest stages. "Who assures us," the naturalists might ask, "that your genealogical trees are anything better than the productions of your imagination?" The students of nature found the relationship of one organic type to another over the course of time their *"quaestio vexata."* Certainly Darwin's theory needed to overcome resistance on this score. Should not philologists' findings be subject to the same critical skepticism?

Schleicher responded by describing the inherent investigative advantage of philology over biology. Lyell had touched on this thesis, yet only in passing; Schleicher made it central to his argument. The comparative philologist or "glossologist," as he called him, enjoyed a surer basis for observation than did the naturalist, for he possessed a fuller record than that found in nature. "To trace the development of new forms from anterior ones is much easier, and can be executed on a larger scale, in the field of speech than in the organisms of plants and animals." As a result, "the glossologist has an advantage over his brother naturalists in this respect. We are actually able to trace directly in many idioms that they have branched off into several languages, dialects, &c."[8]

This abundance of attestation had been made possible by the invention of writing and the survival of written records: had writing not been invented, "the student of language would never have imagined, up to the present day, that such languages as Russian, German, and French, for example, are descended, after all, from one and the same stock." Yet writing *had* been invented, and philologists could therefore "know positively" that certain closely allied dialects had evolved over time into markedly dissimilar tongues.[9]

Schleicher then brought his argument to a striking conclusion. He told Haeckel not only that linguistic study had confirmed Darwin's theory but that philologists were "better off for materials of observation than the other naturalists, and therefore we have forestalled you in the idea of the non-creation of species." One senses that Schleicher enjoyed writing these lines, in which he portrayed comparative philology as the true forerunner of the most advanced school of biological thought. Linguistic kinship, he said, served as a "paradigmatic illustration of the origin of species." Moreover, this paradigm could prove especially helpful "for those fields of inquiry which lack, for the present at least, any similar opportunities of observation."[10] That is, philology could continue to render a kind of aid to those who study natural forms. With these claims, Schleicher gently tweaked his young friend Haeckel, for he exalted his own field at zoology's expense.[11]

These were bold pronouncements. Schleicher was saying that philology led the way by doing what natural history could not, or at least not yet. With its use of old texts, it could demonstrate the actual occurrence of branching descent.[12] When Lyell made his similar point in *Antiquity* about the philologist's advantage, he still admitted that many of the philologist's conclusions rested on indirect evidence.[13] Darwin had been even more modest, for he had implied that linguistic and biological research efforts suffered from a roughly equal lack of evidence even though each produced relatively trustworthy results. For Schleicher, however, the two fields were unequal: he saw philology as supplying Darwinism's lack in the form of positive proofs that transmutation could occur.

Die Darwinsche Theorie und die Sprachwissenschaft was not immediately translated into English. The work became well known in the philological world, yet it would seem reasonable to expect that, striking though its contents were, it would attract little attention among the English-speaking lay public. Nevertheless, it soon got reviewed in the London weekly *Reader,* one of the Victorian era's heroic but short-lived efforts to combine literary and scientific topics in a single periodical.[14] The *Reader* was pro-Darwinian on the science side, and so it provided a fitting platform for the unsigned article describing Schleicher's open letter.

The writer of this piece lamented that the Darwinian naturalists had been unfairly maligned, especially in comparison with philologists. "How is it," he asked, "that the *onus* of proof lies in the realm of Philology, on those who deny, and in that of Natural History, on those who assert, the derivation of the most opposite species from a common parent? Why is the matter-of-course assumption, transported to a region analogous in every respect to its original home, transformed to an audacious theory?" This last sentence followed Schleicher's reasoning explicitly, for it suggested that the "original home" of the descent theory was philology, not biology.

The reviewer's audacious conclusion was that linguistic study should therefore be accorded the higher place among the sciences. "Is it not evident that the case [that being philology] in which the hypothesis becomes a part of our data is that in which science is placed in exceptionally favourable circumstances?" In other words, what the naturalist knew only as hypothesis, the linguistic student knew as fact. The implication was that Darwinism owed an intellectual debt to philology. With this came a rehearsal of Schleicher's central argument, about the advantage the philologist held over the Darwinian naturalist due to the comparative wealth of evidence at the philologist's disposal. Only if the geological record were perfect and complete, and the naturalist's life span reckoned by centuries rather than years, could biological science "stand in that advantageous position with reference to his own subject of investigation which is now the exclusive privilege of the philologist."[15]

The writer predicted, however, that this disparity was only for the short term and that the lag in natural-scientific credibility would soon be made up. Schleicher's pamphlet helped that cause by describing on a small scale what could have occurred in the whole of organic nature. For its part, the *Reader* made this point through a secondary analogy: linguistics was helping biology in the same way that Jupiter and its moons had once aided astronomy. The moons orbiting Jupiter had demonstrated in microcosm the workings of the solar system and, fortunately, had done so at the very time when the principle of gravitation—analogous to the idea of natural selection?—was coming to be understood. By displaying the effects of gravity on an observable scale, Jupiter's moons prepared thinkers to accept that gravity's force could affect an entire system of planets. A similar shift in perspective would occur when scientists and scholars saw how well the Darwinian model of change applied to the observable phenomena of language: it would "impress upon us the probability of its continuance beyond those narrow limits."[16]

The *Reader* thus heralded Schleicher's argument in cocksure fashion, but would these brief columns command much attention? And even if so, what about the far wider attention commanded by Charles Lyell, who had already written on this subject in his *Antiquity of Man?* Would not Lyell's discussion, with its theistic coloring, exert the dominant influence on the Victorian public? True, the published reviews of his book had dismissed its linguistic chapter. Yet Lyell's prestige could not be ignored. It is a measure of the importance attached to these "merely" analogical polemics that the leader among Darwin's younger spokesmen made a point of trying to choke off Lyell's authority in this area.

T. H. Huxley's remarks appeared in the *Natural History Review,* in an article of 1864 examining recent Continental publications on Darwinism. Huxley began by mentioning that a number of such works had come out recently, one of them of particular interest to those who had read Sir Charles Lyell's "remarkable chapter" on language. He referred here to Schleicher's *Darwinsche Theorie,* which, he noted, had independently set forth a language-species parallelism similar to Lyell's. And as he soberly observed, Schleicher's argument carried "all the weight of his special knowledge and established authority as a linguist"—a clear-cut demotion of the geologist Lyell.

Huxley went on to tell the story of how Ernst Haeckel had prompted Schleicher by giving him a copy of *The Origin of Species.* Schleicher's intellectual kinship, he implied, was with Haeckel, already becoming known for his ultra-materialist rendering of Darwinism: he mentioned—by the way, as it were—that Haeckel had expressed this outlook in a "splendid monograph." Finally, Huxley recommended the "excellent notice" of Schleicher's essay that had recently appeared in the *Reader.* Leaving aside these preliminaries and taking up his main topic, Huxley issued a characteristic declaration: "Teleology, as commonly understood, . . . received its deathblow at Mr. Darwin's hands."[17]

Clearly, T. H. Huxley wanted the public to interpret the linguistic analogy within a particular philosophical framework, and the unsigned *Reader* notice contributed to that end. Closely following Schleicher's argument, the *Reader* had embraced the idea that philology confirmed Darwinism, yet it had omitted Lyell's counterpoint about the need for additional "creational laws." Small wonder, then, that Huxley thought it an "excellent notice."

Yet who wrote the *Reader* piece? A roll call of eminent Victorians contributed to that journal, yet nowhere is the authorship of the article on Schleicher indicated. One might suspect Huxley himself, fluent in German and one of the *Reader*'s volunteer editorial assistants from 1863 to 1866.[18] Yet the review's perspective suggests a philologist, not a naturalist.

Evidence points to another known *Reader* contributor: the Anglican churchman, theological writer, classics scholar, and language theorist, Frederic William Farrar (1831–1903). Farrar is perhaps best remembered for his role at the end of Darwin's story, as the clergyman who secured ecclesiastical permission for Darwin's burial in Westminster Abbey. The facts of his career suggest an unlikely ally for the Darwinians, especially the younger ones. Farrar served as a housemaster at Harrow School, headmaster of Marlborough College, chaplain to Queen Victoria, dean of Westminster and, finally, dean of Canterbury. His widest fame would rest on his religious writings, especially his popular *Life of Christ* (1874).

Yet Farrar also had strong scientific interests and even a penchant for naturalistic explanations of human capacities. He was a member of the British Association for the Advancement of Science, through which he would serve alongside Huxley, in the early 1860s, on a committee to encourage better science education in British schools. He especially wanted to see elementary science substituted for Latin verse making, an opinion carrying all the more weight as coming from a classicist and one that Huxley found particularly welcome.[19]

Strangely enough, Farrar was never fully convinced by Darwin's descent theory. He wanted, however, to see it given a fair hearing. He struck up a correspondence with Darwin by sending him his books on linguistic questions, and as a favor, Darwin nominated him for fellowship in the Royal Society, which was granted in recognition of his philological attainments.[20] Assuming that Farrar was responsible for the piece in the *Reader*, we may hypothesize that his authorship came about through a network of personal associations. Haeckel most likely sent Schleicher's pamphlet to Huxley, his chief contact in British science at that time, with hopes that Huxley would show it to Darwin. Huxley probably recruited Farrar to write the review.

On the Origin of Languages and Dialects

Throughout the 1860s, the intellectual relationship between F. W. Farrar and F. Max Müller played a significant, yet thus far mostly overlooked, role in

Darwinian debate. The most cordial of friends, Farrar and Müller stood at odds on major linguistic questions. One thing they agreed upon, however, was the scientific status of philology. Both wanted to distance their field from its background in antiquarian and literary pursuits and to move it closer in spirit to natural science. Much of philology, especially the study of the classical European languages, was still a text-centered, humanistic discipline. Müller and Farrar, however, embraced the exacting comparative-grammatical methods of German *Sprachwissenschaft*.

This new philology, rhetorically linked to Cuvier's comparative anatomy and now routinely (at least in Britain) compared with Lyellian geology, saw itself as a science, one deserving admission into the circle of the most advanced research-oriented disciplines. We have seen this thesis promoted already in August Schleicher's pamphlet. Outside Germany, however, these aspirations continued to be touched with strains of insecurity, and many Anglophone philologists saw their field as still in need of legitimation. Said one knowledgeable writer in the early 1860s, "*Philology* has, to some ears, a slightly *unscientific* association." Max Müller addressed the problem forcefully at London's Royal Institution, when he assured the audience of his 1861 lecture series that there "really is such a thing as a *science* of language."[21] Müller said such things often, and Farrar argued pretty much the same case. We will see as we go the importance of these aspirations, that is, how they became linked analogically to the controversy over Darwinism.

Farrar and Max Müller did battle with each other on two main fronts, the more obvious of these by far being the question of the origin of language. Surprisingly, perhaps, the clergyman Farrar was one of Victorian England's two leading exponents, the other being Hensleigh Wedgwood, of a naturalistic explanation of the beginnings of human speech.[22] Both writers held that articulate utterance had begun with the conscious mimicking of sounds occurring in nature. Farrar's argument first appeared in his *Origin of Language* of 1860. Even though this work was published just after Darwin's *Origin of Species*, Farrar does not seem to have written it out of any desire to buttress the idea of human evolution. Rather, as he said in the book's preface, he had been inspired by his own researches on the topic at hand and by ideas borrowed from the French savant Ernst Renan.[23] In any case, Max Müller's treatment of the origin of speech in his Royal Institution lectures responded in large part to Farrar's statement. Müller did not need to name his target when he derided the bow-wow theory of language.

Müller aimed his lectures, however, not only at Farrar but in a larger sense at

Darwin. Darwin noted this himself in his correspondence with Asa Gray, when he said that Müller's book contained "covert sneers directed at me, though he seems to get the better of this toward the end."[24] The remarks Darwin referred to here most likely had nothing to do with Müller's mystical theory of the origin of language. Darwin of course found that theory most unsatisfactory. Yet Müller addressed the problem of origins only in his final lecture, and Darwin said that, by this point, Müller had stopped sneering.

I suspect that Darwin's complaint actually referred to Müller's stated rejection of transmutationism. This theme appeared in the opening lecture, in connection with Müller's argument that the study of language was not a historical discipline but was one of the physical or natural sciences. Müller was aware of the resistance this quixotic thesis would provoke, and he anticipated various objections. One was the obvious fact that languages undergo change. Taking the part of a critic, he declared it "a well-known fact, which recent researches have not shaken, that nature is incapable of progress or improvement."

As Müller thus presented the problem, his case for linguistics' status as a natural science was obliged to conquer his own sincere effort to vindicate a pre-Darwinian view of species. Müller held that every part of nature—mineral, plant, and animal—remained "the same in kind from the beginning to the end of its existence." Variable linguistic phenomena would therefore appear to be excluded from the natural sciences. For this problem Müller had an ingenious solution: although languages underwent continual change, they did not have histories as the term was then usually understood, for their changes were not brought about by the conscious intent of individuals. This answer is intriguing in itself, although the important point for our purpose is the way Müller set up the question.[25] He affirmed a static, non-Darwinian view of organic life—this despite "recent researches." Darwin no doubt referred to this passage when he complained of Müller's "covert sneers."

Admittedly, Max Müller's position on Darwinism is hard to pin down. His eloquence came in such abundance that it tended to cover both sides of a question at once, and the higher synthesis he offered was seldom manifest in any one statement. Müller later suggested that his quarrel with Darwinism concerned human descent only, especially after the appearance of *The Descent of Man*.[26] He hedged his bets all along, moreover, by showering Darwin with praise and by using Darwinian concepts in his exposition of linguistic themes, for example, the "struggle for existence" among superabundant word roots discussed in his first volume of lectures. His second volume extended the application of the selection idea forward in human history, to the long-term growth of linguistic and cultural forms; in this instance, Müller's argument displayed not only literary elegance but also impressive sociological acumen.[27] Yet Müller kept his analogizing within strict bounds: while he applied Darwinian categories to language, he did not use linguistic themes to illustrate or support Darwinism. That is, he never used the analogy in the same way as Darwin and his defenders did.

Indeed, what especially concerns us is a subtle anti-Darwinian analogy embedded in Max Müller's linguistic science. It is an aspect of his work easy to overlook, for it constituted a subordinate theme, and only an implicit one at that. It also ran against—directly against—the message of Müller's comparative philology described in chapter 3. There we saw his writings inspiring Charles Lyell's *Antiquity* analogies and inducing Asa Gray to declare that the idea of descent could do good service in Darwin's cause. Neither Lyell nor Gray—nor Darwin, it seems—noticed the opposing message insinuated in Müller's descriptions of the history of language.

But someone else noticed: F. W. Farrar. Darwinism-by-analogy formed the second linguistic front on which he and Müller battled each other, creating an issue less conspicuous yet historically more significant than their debate over the origin of speech. Paradoxically, although he never became an evolutionist, Farrar made greater claims than did any other British writer, Lyell included, for the linguistic analogy in support of Darwin's theory. We have already seen the line taken in the *Reader* article commending August Schleicher's outlook; Farrar's pro-Darwinism would appear even more frequently in his polemics against Müller.

Farrar and Müller did have this in common: because they held roughly similar religious outlooks, questions of human transcendence generally did not enter their analogical disagreement. In this respect, the debate between these two scholars constituted a straight up-or-down vote on the plausibility of biological common descent, apart from what it implied about man's place in nature. Their mode of debate, however, was far from simple. Master polemicists, they each fashioned texts having multiple and partly hidden levels of meaning. Their shared discourse also cut a serpentine path, for each shifted perspective at times to fit the demands of the occasion. Yet by the eve of Darwin's *Descent of Man*, Müller's and Farrar's positions would at last be stated openly and the interpretations of their linguistic parables made plain. Working from the beginning of the 1860s up to that terminus, I attempt to lay bare this long-obscured roadbed of Darwinian debate.

Müller's intent to address biological questions by analogy can only be inferred at its beginning, and his viewpoint arose at least in part from purely linguistic considerations. Even so, his writings during the 1860s betrayed a new and unannounced yet clearly antievolutionist polemic. Prior to this, Müller had been the first really to popularize in England the idea of the common descent of related languages and parallel myths, a thesis that made an early appearance in his 1851 review of the English translation of Franz Bopp's *Comparative Grammar.* There Müller spoke of "tongues which sprang from a common source," includ- ing "the Sanskrit, with all the different dialects, which have sprung from it."[28] Müller then presented this thesis with particular clarity at the outset of his *Comparative Mythology* (1856), where he illustrated the idea of an Aryan linguis- tic family by showing the same genealogical pattern on a smaller scale in the Romance languages.

At about the time when Darwin's *Origin of Species* appeared, however, newer findings in Romance philology led Müller to adjust his perspective on the genealogical tree.[29] In his 1861 lectures, he called attention not only to the tree's spreading branches but also, in separate passages, to an opposite pattern: the converging roots that fed the Romance languages from beneath. Rather than view dialects as descendants of a given parent tongue, Müller now regarded these as tributaries, in effect turning the family tree upside down. Unified languages, he argued, emerge only at the latter end of a long process, usually because of the consolidating influence of literary cultivation. He gave the example of "Germanic," the conjectured parent of German, Dutch, Icelandic, and the several Scandinavian languages. Müller regarded this prototongue as a mere abstraction, something that had never had a real historical existence; the reality was a large number of contiguous local dialects gradually coalescing into a few distinct national tongues.[30] Similarly, although the six Romance languages had derived in a general sense from Latin, classical Latin alone did not supply a full accounting of their origins. As Müller noted, the real feeders of these so-called daughter languages were "the vulgar, local, provincial dialects of the middle, the lower, and the lowest classes of the Roman Empire." Müller made the point more plainly several years later, when he remarked on "the limited applicability of a genealogical classification" of languages.[31]

This new theory arose from purely philological considerations, did it not? Müller had legitimate reasons for making this case for the original diversity of dialects (although his use of that term in this regard might be disputed), and it clearly was something he sincerely believed. Yet he would also apply this thesis outside the strictly linguistic realm. It was probably no accident that the first clear statement of Müller's dialects theory came just after George H. Lewes's "Studies in Animal Life" appeared in 1860. There, as we saw, Lewes quoted at length the demonstration of Aryan linguistic kinship from the opening of

Müller's Oxford lecture, and he used this thesis to recommend Darwin's newly published transmutation hypothesis. Was Max Müller pleased with the favorable publicity thus given his work? The larger story suggests that he was not.

An Indirect Debate

The first half of the 1860s saw only the barest beginnings of F. W. Farrar's challenge to Max Müller on the analogic front. The *Reader*'s review of August Schleicher's pamphlet may have served this function to some extent. Then came the *Reader*'s polite notice, again probably penned by Farrar, of Max Müller's second London lecture series. The reviewer made a point of casting Müller's views in Darwinian-sounding phrases: the new volume's consistent theme was "the investigation of the process by which linguistic variations have come about, supplemented by a discussion of the results which the process of mutation has itself produced in human thought."[32]

Farrar's much more important work at this time, however, was his *Chapters on Language* (1865), a revised and expanded version of his *Origin of Language*. Here he answered various critics, Müller among them, of the imitative theory of speech. He also included this tangential yet highly significant revelation: "Disbelieving, on the scientific ground of the Fixity of Type, the Darwinian hypothesis," he nevertheless found it "neither irreverent nor absurd." Farrar therefore urged his readers not to dismiss that hypothesis out of hand.[33] Thus, for the first time, Farrar openly expressed his dissent from Darwinism yet also his opinion that it should be given a fair trial.

Publication of *Chapters on Language* brought the beginning of Farrar's relationship with Charles Darwin. Farrar sent a copy of his new book to Down House and received an enthusiastic reply. "As I have never studied the science of language it may perhaps be presumptuous, but I cannot resist the pleasure of telling you what interest and pleasure I have derived from hearing read aloud your volume." Darwin especially appreciated Farrar's choice of enemies on the subject of the origin of speech. "I formerly read Max Müller, and thought his theory (if it deserves to be called so) both obscure and weak; and now, after hearing what you say, I feel sure that this is the case, and that your case will ultimately triumph." Darwin added: "My indirect interest in your book has been increased from Mr. Hensleigh Wedgwood, whom you often quote, being my brother-in-law."[34]

Then, as to the remark Farrar had made about his hypothesis, Darwin said this: "No one could dissent from my views on the modification of species with more courtesy than you do. But from the tenor of your mind, I feel an entire and comfortable conviction (and which cannot possibly be disturbed), that, if your studies led you to attend much to general questions of Natural History, you would come to the same conclusions that I have done." He asked whether Farrar had read Huxley's "little book of Six Lectures" and offered to send him a copy. Then he took the opportunity to play upon Farrar's linguistic imagination, probably not suspecting that Farrar himself must have been the writer who had discussed Schleicher's *Darwinsche Theorie* in the *Reader*. What Darwin did recall was the hypothetical case, in Lyell's *Antiquity*, of an illiterate tribe's conclusions concerning the mutability of language. "Considering what geology teaches us, the argument for the supposed immutability of specific types seems to me much the same as if, in a nation which had no old writings, some wise old savage was to say that his language had never changed; but my metaphor is too long to fill up. Pray believe me, dear sir, yours very sincerely obliged [etc.]."[35]

Writing in response, Farrar gave no reaction to the unfinished metaphor, although he surely saw its point. Yet he did give a surprising explanation of his own resistance to the transmutation hypothesis. Farrar apparently had no qualms about the nonteleological character of natural selection, or even about the evolutionary emergence of humankind. Rather, he said that his commitment to the doctrine of fixity of type was rooted in his beliefs concerning human ethnology: "Perhaps my Polygenist prejudices may have led me to it."

Having already read the "lectures to workingmen" that Darwin recommended, Farrar knew that Huxley believed in monogenesis, and he assumed that Darwin did as well. "Now I confess that, so far as I can see, History, and even Tradition, as far back as their primeval dawn, *prove* to us the existence of the several human races unchanged from their present physical characteristics; . . . Is not this a very strong argument for the Polygenist?" Farrar asked.[36] He of course knew that polygenism was not "absolutely *incompatible*" with Darwin's view of species. Yet he discerned that the two viewpoints seemed to exclude each other. He did not spell out his reasoning on this point but, clearly, it depended heavily on analogy. In any case, Farrar ended by declaring his intellectual neutrality: "My state of mind, however, on this question is a mere *suspension of assent,* and nothing would surprise me less than the discovery of some fresh palaeontological fact, or the artificial production of some new species, which would practically decide the question in favour of your hypothesis."[37]

Farrar made his neutrality manifest several years later, when he argued, indirectly, both sides of the evolution question back to back. First he took the skeptical position. He questioned the notion that the main grammatical types of the world's languages—isolating, agglutinating, inflective, and polysynthetic—represented successive stages of linguistic development. This stage theory extended Wilhelm von Humboldt's classification of languages according to the complexity of the typical word structure in each. In isolating languages, Chinese being the classic example, most words were indissoluble monosyllables. Agglutinating tongues added a class of words that had been "welded" from independent roots yet with at least one element retaining its original significance. Inflective languages, the Indo-European and Semitic families, included words of which no part remained identifiable as an independent root.[38] A fourth type, polysynthetic (also called holophrastic), applied mainly to the American Indian languages, in which "words" expressive of an entire cluster of ideas were formed from strings of dependent syllables.

According to the theory of successive developmental stages, every inflective language emerged from an agglutinative ancestor, and every agglutinative language was built on a monosyllabic foundation. Many nineteenth-century philologists embraced this theory. Max Müller affirmed it, especially in his 1868 Rede Lecture at Cambridge University, "On the Stratification of Language." "At one time Sanskrit was like unto Chinese, and Hebrew no better than Malay." This "transition from one stage to another is constantly taking place under our very eyes." After all, the stages were not absolutely distinct. Since the inflectional languages grew from a previous, noninflecting stage, certain transitional forms survived. "Even Chinese is not free from agglutinative forms, and the more highly developed among the agglutinative languages show the clearest traces of incipient inflection."[39] This was a purely linguistic thesis for Müller, with no intended reference to Darwinism. Indeed had it been otherwise, the analogy would have been proevolutionist.

Farrar, however, turned the philological issue into something more significant. Twice within the space of a few months, in the same year as Müller's Rede Lecture, he questioned whether the Aryan and Semitic tongues had indeed ascended through the lower strata of grammar formation. Especially telling was the way Farrar echoed a stock theme of anti-Darwinian skepticism. He conceded that there might be warrant on purely theoretical grounds for the notion of transition from stage to stage. But, he declared,

> there is not the faintest shadow of *evidence* to prove that any monosyllabic, agglutinative, holophrastic, or synthetic language has ever at any previous period been otherwise than monosyllabic, agglutinative, holophrastic, or synthetic. . . . The facts of historical philology make it *almost* impossible to suppose, and *quite* impossible to prove, that any language of any group has ever passed, or can ever pass by stages however slow, into another group.[40]

Farrar addressed these comments to the members of the Cambridge Philological Society and repeated them in a review of Müller's *Stratification*. He dealt on the surface with linguistic phenomena only, although one suspects that at least some in his audience would have seen the connection with Darwinism. Farrar himself later made that connection explicit.[41]

Yet Farrar had said that he was open-minded, and meanwhile he made good on this. Within a few months, *Macmillan's Magazine* carried his "Philology as One of the Sciences." As the title suggested, Farrar wanted to vindicate philology's status as among the truly scientific disciplines. A general impression of the field's link to natural science, he said, could be gotten from the many natural-historical "analogies and illustrations" used in linguistic theory: these cropped up "spontaneously, and almost unconsciously" in nearly every recent work on that subject. Much more important, however, was the similar method shared by philology and the sciences. Echoing Auguste Comte's schema of three stages in the development of the sciences, Farrar said that philology had already passed through its theological and empiric stages and had entered the final positive stage. (In illustrating the first of these, Farrar disparaged "the theological philology of M. de Bonald"; French and Catholic, Bonald was fair game for abuse and hence a useful surrogate for Max Müller.) If botany, for instance, were considered a full-fledged science, then philology should be as well.[42]

Farrar elaborated particularly on this parallelism between philology and botany. Each field had established its scientific credentials by uncovering the true similarities and differences among its own phenomena, and these distinctions showed the way to a natural system of classification. "A botanist who was a mere corollist would not have been likely to class in the same natural order of *Ranunculaceae,* plants so externally dissimilar as larkspur, columbine, and buttercup." To this he likened the comparisons made by the linguistic scholar, who saw that widely distinct words often should be classed together. "Who, for instance, would think of comparing the Gothic *faihu,* 'cattle,' with the Latin *pecus,* if his etymology were founded on mere appearances?"[43] Equivocally hinting at a non-Darwinian ideal archetype as much as an actual ancestor, Farrar drew another simile: "Once more, exactly as the botanist assumes a certain ideal symmetry, even when every species of a family deviates from it in one or other particular, so the philologist often assumes a primordial form which alone explains its divergent derivatives." Intermediate forms needed to be posited as well: "Nor could he [the philologist] without the intervention of many varying forms conjecture the identity of the words *five* and *quinque.*"

There was a considerable element of caution in these remarks. Farrar chose botany as an example, for the classification of plants was less controversial than that of animals, among which man might be included. He also cited individual etymologies rather than the entire tree of Indo-European languages, even though the latter would have made the more strikingly Darwinian comparison. In another sense, however, these parallelisms were bold. They actually went beyond mere analogies, in the sense of illustrative similarities recognized after the fact. Rather, as Farrar argued, they showed a commonality of research procedure that philology and the biological sciences employed from the outset.

Preliminary conclusions may now be drawn about Farrar's use of linguistic analogies, both for and against Darwinism. On the one hand, his skepticism about Darwin's theory was more thoroughgoing than that seen in Charles Lyell's *Antiquity* chapter. Lyell doubted the efficacy of natural selection and posited new creational laws to explain transmutation. Farrar doubted transmutation itself. On the other hand, Farrar did not betray the level of ambivalence toward Darwinism suggested in Lyell's chapter, in part because he introduced his pro- and anti-Darwinian analogies in separate writings. And on the whole, the former would outweigh the latter. At the close of his "Philology as One of the Sciences," Farrar mentioned Darwinism for the first time since his book *Chapters on Language:* "I had intended to show the important light which the science of language may be proved to throw upon the Darwinian theory; but space forbids me, for the present."[44] Implicitly, of course, he had just shown that very thing. Yet as he hinted, he still had more to say about linguistics in relation to Darwinism, and before long he would do so.

The year of Farrar's indirect pro-and-con over Darwinism, 1868, brought the untimely death of August Schleicher, thus ending the pioneering era of German-led comparative philology. Yet this event also inspired the first English translation of Schleicher's pamphlet *Die Darwinsche Theorie und die Sprachwissenschaft,* which bore a provocative new title: "Darwinism Tested by the Science of Language" (1869).[45] What would be the response now that Schleicher's views were given wider exposure in England?

The *Athenaeum* dismissed the work, partly because of its title. "As a test of Darwinism, it is almost laughable. It is no *test* at all of Mr. Darwin's theory in relation to a tolerably sufficient field of observation." The analogy itself the reviewer described as "an ingenious but thoroughly forced parallel between the leading features of Mr. Darwin's theory and the history and derivation of languages—amusing for its ingenuity, but utterly unconvincing as an argument for Darwinism."[46] Even the Utilitarian *Westminster Review* had doubts. The scientific writer William Sweetland Dallas predicted that Schleicher's analogy would have a "beneficial influence on the study of language; but the opponents of Darwinism will hardly accept his arguments as supporting the general theory."[47]

Despite these reactions, the English edition of Schleicher's pamphlet launched a new phase in the career of the language-species analogy. At least two prominent writers took Schleicher's arguments quite seriously, and they expressed their opinions in a new and auspicious setting. The weekly journal *Nature*, founded late in 1869, marked an epoch in modern scientific publishing. Many of the young stars of Victorian science contributed: Huxley, Hooker, John Tyndall, Herbert Spencer, and others. Most of these were members of the X Club, a Darwinian clique within the Royal Society, the British Association, and the Royal Institution. *Nature* was blessed with other friends as well, including the circle of writers who sustained another newly established London weekly, the *Academy*.[48] Yet in one instance, the blessing turned into a curse, for Max Müller was the *Academy*'s authority on philological matters. One wonders how the friends of *Nature* reacted when they read what Müller contributed in January 1870.

Apparently, the British readership's sudden access to Schleicher's views impelled Müller to make his own position on the relationship between language and Darwinism more explicit. Müller's review of Schleicher's translated pamphlet was an astounding performance, for it distilled in a few columns a host of polemical themes.[49] Like the reviewer for the *Athenaeum*, Müller objected first of all to the translation's title. "Professor Schleicher could hardly have thought that the truth or falsehood of Mr. Darwin's theories depended on any test that can be applied to them by the Science of Language. But he thinks rightly that the genesis of species, as explained by Mr. Darwin, receives a striking illustration in the genealogical system of languages."

In this limited sense, Müller conceded, the historical pattern of development seen in the Aryan and Semitic tongues "may be useful as a kind of confirmation of Mr. Darwin's theory." Again, he appeared to affirm the comparison: "No reader of Mr. Darwin's books can fail to see that an analogous process pervades the growth of a new species of language, and of new species of animal and vegetable life." Yet what Müller actually meant by these remarks came out only at the end of his review. Meanwhile he had other complaints: chiefly, he asserted that the selection process in the linguistic realm was not blind but rational in character, "for what seems at first sight mere accident in the dropping of old and the rising of new words, can be shown in most cases to be due to intelligible and generally valid reasons."[50] That is, linguistic adaptation was teleological.

The argument in Max Müller's *Nature* piece managed to sound strikingly

Darwinian even as it created an ingenious travesty of Darwin's descent theory. Here, at long last, Müller made plain his oppositional version of the language-species analogy. His themes on the linguistic side are already familiar: the struggle for existence among rival words and grammatical forms within a given language, and the transition from scattered dialects toward unified tongues. At each of these levels, both within languages and among them, Müller saw diversity giving way to unity. In *Nature,* however, Müller did something new: he conflated these two processes and argued that the result was reflected in nature. At the level of competing word forms, the pattern suggested an apt analogy with Darwinian selection, unquestionably a kind of winnowing procedure. Yet at the level of development among related languages, Müller's analogy denied the ramifying character of biological descent.

Müller introduced this argument through insidious means, ostensibly faulting the philologist's rather than the naturalist's understanding of evolution. "There is one point on which Professor Schleicher seems to have misapprehended the meaning of Mr. Darwin. According to him [Schleicher], the different species of the Aryan as well as of the Semitic languages presuppose each a typical language from which they are genealogically derived." Yet this typical prototongue was indeed that, Müller argued; it was only an ideal construct, not intended to represent a real common ancestor. Now, more plainly than at any time before, Müller overturned his own branching-genealogical argument, made so convincingly in the "Comparative Mythology" lecture of 1856. He now flatly declared, "We should not attempt to derive the great dialects—viz. Greek, Latin, Celtic, Teutonic, and Slavonic—from a presupposed primitive Palaeo-Aryan type of speech."[51]

Against this backdrop, Müller spelled out exactly what his dialects theory represented in the biological realm.

> In tracing the origin of species, whether among plants or animals, we do not begin with one perfect type of which all succeeding forms are simply modifications, but we begin with an infinite variety of attempts, out of which by the slow but incessant progress of natural selection, more and more perfect types are gradually elaborated. . . . It is the same with languages. The natural state of language consists of unlimited dialectic variety, out of which, by incessant weeding, more and more definite forms of languages are selected.[52]

Müller thus collapsed together two wholly distinct levels of Darwin's theory by basing the larger descent pattern on the selection principle. Although different species sometimes competed with one another for food and territory, resulting in the triumph of some species over others, this is not what the *Origin* meant by selection; natural selection, rather, was based on competition among individuals of the same species.[53] Ignoring this distinction, Müller garbled the *Origin*'s descent theory so as to make it comport with his upended linguistic tree. Perhaps he wanted to show that his own reading of Darwin's descent idea was more plausible than the "orthodox" version, that a struggle for existence among species would produce paring down rather than branching out. Perhaps also, Müller wanted to imply that the branching-genealogical schema, both in philology and in natural history, was woefully infected with typological thinking; for, as the above passages show, he equated evolutionary ancestors with ideal and perfect archetypes.[54]

Not surprisingly, the editors of *Nature* welcomed a second opinion on August Schleicher and the language-species comparison; not surprisingly, the author of that opinion was F. W. Farrar. As a result, within just a few months of its founding, Britain's preeminent natural-scientific journal printed not one but two articles on language. Both of these, moreover, would provide grist for Charles Darwin's further imaginative polemics.

The second article in *Nature* began with a tactful preface. "It struck me that many readers might be glad to have some further account of Schleicher's views."[55] Appropriately enough, Farrar emphasized comparative philology's picture of branching descent, thereby countering Müller's emphasis on dialects coalescing into languages. And he again promoted the advantage thesis, in this case by pointing out the difference between the family tree diagrams found in Darwin's *Origin* and in Schleicher's pamphlet.

> Here the philologist has a distinct advantage, and the study of his results may be most suggestive to the naturalist: for the Darwinian diagram is to a great extent ideal and hypothetical; while the table of languages is merely an expression of indisputable discoveries. Any one who has clearly understood the certainty of the fact, that languages at first sight so different as Greek and French, Icelandic and Portuguese, Sanskrit and Lithuanian, are yet connected with each other by close bonds of union, and that the phenomena they exhibit are due to gradual differentiation from a single stock, will undoubtedly be more able to conceive the possibility of Newfoundlands, and Greyhounds, and King Charles's Spaniels, and Wolves being lineal representatives of a common type.[56]

As this passage shows, Farrar no longer restricted himself to etymological genealogies; rather, he built his analogy on the larger thesis of common descent among the Indo-European languages.[57]

More than this, Farrar's version of the advantage thesis was bolder than anything that had yet appeared in Britain. In Lyell's metaphor in *The Origin of Species,* and in the first half of Lyell's own language chapter in *Antiquity,* linguistic analogies had served to vindicate gradual transmutation in spite of the missing pages in the paleontological record—a thesis consistent with Lyell's and Darwin's tendency to focus on the mere absence of contravening proof. Following Schleicher, Farrar made a much more extravagant claim: the philologist could "not only *prove*, where the naturalist must be content to *conjecture*, but can also more easily exemplify the birth of new forms out of anterior ones, and can carry out his examination on a greater scale."[58] Hence Farrar was able to regard Darwinism as both tested and proved by the science of language.

In closing, Farrar said that he had left his discussion incomplete, as he had omitted certain linguistic facts "which might be alleged with great force in favor of an opposite view"—that is, an anti-Darwinian perspective. He did not elaborate but referred readers to his Cambridge lecture rejecting the notion of transitions between the isolating, agglutinative, and inflecting language types.[59] Yet by relegating this side of the argument to a mere citation, Farrar gave the impression that he viewed Darwinian descent with the greater favor.

Here, again, F. W. Farrar set Schleicher's version of the linguistic analogy before the British public. And *Nature* was not the only forum in which he did this. He had only recently presented these same ideas in his lecture series "The Families of Speech," delivered at the Royal Institution in London. This event brought the long-running duel full circle, for Max Müller had established his formidable popular reputation at that same site eight years earlier. Since the middle of the decade, however, the Royal Institution had come to be dominated by the X Club, and that influential clique no doubt had a hand in inviting Farrar to deliver his own lectures on language. T. H. Huxley, for one, would have seen the advantages of having an eminent cleric and an accomplished philologist present views more congenial than Müller's.[60]

The published version of *Families of Speech* (1870) brought out the desired contrast. Obviously inspired by Schleicher, Farrar scattered linguistic tree diagrams throughout his text to illustrate the branches and affinities of the Indo-European tongues (see figs. 4.2, 4.3). He also confronted the issue of Darwinism directly, although with considerable irony. Max Müller's 1861 lectures had included "covert sneers" aimed at *The Origin of Species*. Farrar now remarked to a similar London audience:

> We are all of us old enough to remember the burst of ignorant derision and theological contempt with which the majority of unscientific Englishmen greeted the announcement of the Darwinian hypothesis. . . . Now the first announcement of the Aryan unity was received with a large amount of similar incredulity. . . . But in spite of all this doubt and ridicule, Science . . . quietly and unconcernedly wins its way. What was at first the bold and brilliant conjecture from Sir William Jones, has now been proved, by half a century of magnificent and incessant labours, to be an

unquestionable fact. Fifty years ago, few would have believed that Dutch, and Russian, and Icelandic, and Greek, and Latin, and Persian, and Mahratti, and French, and English, were all indubitable developments from one and the same original tongue, and that the common ancestors of the nations who speak them were . . . living together as an undivided family in the same pastoral tents. In the present day, no one doubts the fact, except a few intrepid theologians.[61]

This was a picture of comparative philology establishing its credibility in the court of scientific opinion. The point, of course, was to counsel suspension of further hostility toward Darwinism, in light of the fact that similarly new and suspect theories had been proved right in the recent past.

The most striking feature of both Farrar's *Nature* article and his *Families of Speech* is the way the linguistic analogy now turned back on itself. Like both Schleicher and Lyell, Farrar spent much of his time describing the application of Darwinian categories to language. Yet he did this in a new way, one that blurred the two possible directions in which the analogy could point. Indeed, by closely interweaving the organic and linguistic metaphors, he nearly resolved the two sides of the analogy into a single narrative.

> [Philology] can throw light on one of the most important problems of science by showing in actual process before our eyes, the origin of linguistic species from a single genus; she can, with an almost infallible certainty, and with a skill not inferior to that of the comparative anatomist, reconstruct extinct and archetypal forms of language by the comparison of divergent yet closely-related dialects; by examining a speech subjected to foreign influences she can strikingly exemplify the phenomena of hybridism; pointing to an immense number of languages widely separated and mutually unintelligible, and which have existed in their present condition as far back as history can reach, she can yet prove that these species are *not* primitive; . . . and she can show further the enormous influence which, without any sudden changes or violent catastrophes, can be exerted in the progress of centuries, by this continuous differentiation.[62]

In short, Farrar stressed how Darwin's hypothesis could "not only be abundantly illustrated, but positively confirmed, by the researches of the philologian into dead and existing tongues."[63] Yet this produced a nearly circular argument, for, as he described it, philology was able to confirm Darwinism in part because Darwinism seemed to explain so well the evolution of language. This reasoning was similar to Charles Lyell's in *Antiquity,* for it presented the analogy cutting in both directions. Unlike Lyell, however, Farrar made these two theses appear simultaneously, without a transition from the one to the other. This collapsing of the two sides of the analogy was only appropriate since Farrar here omitted the clever arguing on both sides of the question in which Lyell had engaged.

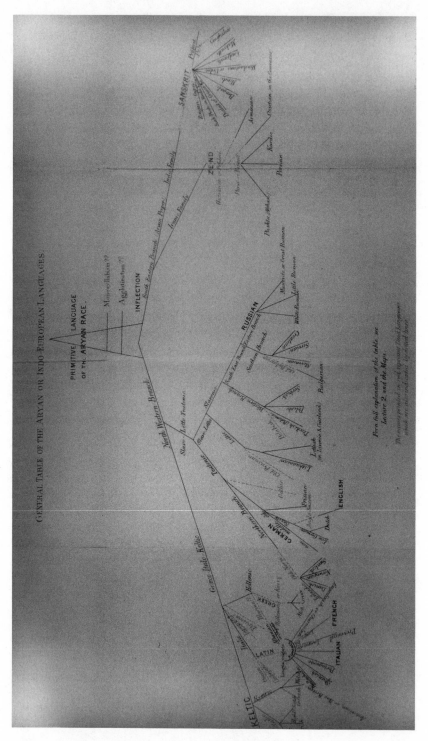

Fig. 4.2. F. W. Farrar, genealogy of Indo-European languages from *Families of Speech* (1870).

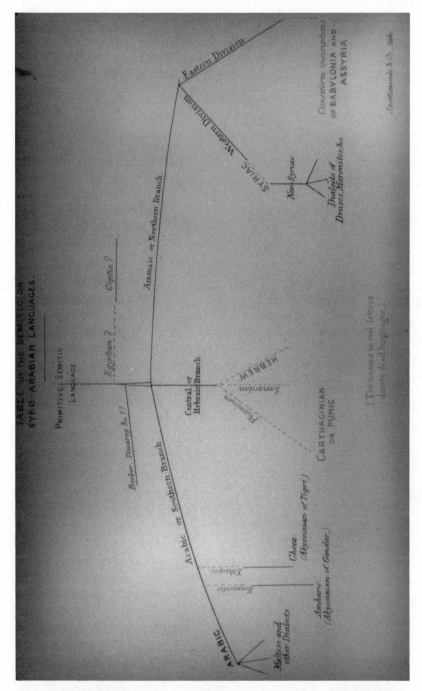

Fig. 4.3. F. W. Farrar, genealogy of Syro-Arabian languages from *Families of Speech* (1870).

At this point a summary of the analogic debate between F. W. Farrar and F. Max Müller may be useful. Even after midcentury, linguistic study could still be considered a fledgling field, and its spokesmen, especially those outside Germany, were still trying to establish its reputation as a science. Farrar and Müller were equally interested in furthering this end. Müller made the most audacious bid for this status. As he announced at the beginning of his second series of Royal Institution lectures: "*One* thing I feel more strongly than ever,—namely, that, without the Science of Language, the circle of the physical sciences, to which this Institution is more specifically dedicated, would be incomplete."[64]

Furthermore, both Müller and Farrar tried to attach their field to the avantgarde of biological science, to hitch their wagon, at least in some sense, to Darwinism's rising star. Müller pointed to the analogic relationship between Darwinism and linguistic phenomena, yet confined himself to the claim that Darwinism was philology's conceptual benefactor: he seized upon natural selection as an explanatory model for the origin of language and for long-term linguistic and cultural development. Farrar's two-way use of the analogy, with its additional pro-Darwinian dimension, was even more promotional of his own field: by supporting the claims of Darwinism in this fashion, Farrar thrust the advantages of philology into the limelight as well. This pro-Darwinian boldness, then, stemmed as much from extrinsic commitments as from a desire that Darwinism be given a fair hearing. Following Schleicher's playful analogic finesse—yet arguing in deadly earnest—Farrar made the plausibility of Darwin's descent theory and the legitimation of his own field go hand in hand.

Always cool toward Darwinism, London's *Athenaeum* had been openly hostile to Schleicher's pamphlet and so could be counted on to disparage Farrar's *Families of Speech*. Tellingly, its review mentioned a familiar-sounding notion about dialects. "In the otherwise excellent table of the Indo-European languages, Mr. Farrar by a slip has marked all the Romance languages as derived from the Latin; whilst the other dialects of Italy, which undoubtedly contributed much to their formation, are allowed no share in the work; and a similar oversight occurs in the Indian family."[65] Hence the analogic element, the dispute over what linguistic study suggested about organic species, still floated beneath the surface of the text.

The Threat from Transcendentalist Morphology

The ultimate goal of this chapter is to account for an anomalous feature of Darwin's *Descent of Man and Selection in Relation to Sex*. Here Darwin at last presented his case for the evolutionary emergence of humankind; to do this, he needed to demonstrate a fundamental similarity between humanity and the higher mammals in their mental faculties, including the faculty of speech. This was an argument about language itself, and the introduction of linguistic analogies in this context might appear irrelevant, perhaps even tendentious. Darwin's discussion contained, nonetheless, a series of comparisons between linguistic and biological evolution. These analogies were more elaborate and revealed a much more detailed knowledge of philology than did his illustrations in *The Origin of Species*. And although compressed within a single paragraph, they rivaled in sophistication the full chapter composed in this mode in Lyell's *Antiquity of Man*. Yet Darwin did not respond to Lyell's chapter only, but to F. W. Farrar's and Max Müller's writings as well. And this was not all, for still other influences came into play. Before examining the *Descent* passage, one must look at these further aspects of the design-in-nature argument in midcentury Britain.

By 1864, Darwin and Asa Gray had reached an impasse in their discussions of supernatural design; a symptom was that Darwin now stopped advertising the reprint of Gray's *Atlantic Monthly* articles in new editions of the *Origin*.[66] At this same time, however, the zoological paleontologist Richard Owen presented a new threat to Darwin's philosophy of nature. Owen's relation to Darwinism was and still is notoriously hard to pin down. Probably an evolutionist in the general sense, Owen did much to prepare the way for Darwin's theory. Yet in the years after the *Origin* appeared, he took an active role in the opposition.

Owen had long promoted transcendentalist comparative anatomy, which stressed formal structure rather than functional adaptation; he thus rejected the utilitarian physiology of Cuvier and Paley. He especially championed the idea of homology, a notion he perfected during the 1840s and 1850s in his research on vertebrate morphology. Homologies were similar structures in unlike animals, such as the bone configuration shared by the human hand, the bat's wing, and the whale's flipper: all of these were formed on a single ideal plan or "archetype," despite their very different conditions of use. That is, they betrayed as a group a minimum of differentiating adaptation. (The converse of morphological homologies were analogies, organs having similar function but different structure.) Design in nature could thus be regarded as conformity to type, revealing the imprint of the divine intelligence in the structures of organisms. T. H. Huxley derided this notion, what he called "the Platonist coloring of the vertebrate blue-print."[67]

In a bold and imaginative stoke, Charles Darwin refashioned Owen's vertebrate work during the 1850s, so as to make it fit his own theory. He did this by converting the ideal archetype into a real ancestor, by accounting for homologies by community of descent, and by turning the successive introduction of typically related forms into gradual transmutation. Yet Darwin went on to demonize Owen unfairly during the 1860s, portraying him, with help from Huxley, as a benighted creationist. It is true that Owen doubted the evolutionary emergence of humans. He was also skeptical about natural selection, for he denied the reign of chance in nature. Like Lyell and Gray, he believed in guided evolution, a theme he expounded, however obliquely, in his 1858 presidential address to the British Association: there he spoke of the "continuous operation of Creative power" in the ongoing development of life.[68]

A popularized version of Owen's views appeared in the writings of George Campbell, the eighth duke of Argyll, especially in his immensely popular *Reign of Law* (1866). (This book went through five editions in two years.) Like other critics before him, Argyll noted that natural selection failed to explain the emergence of variations; it merely suggested how they were preserved once they had appeared. The only true guide to nature, then, was its manifest plan, the outward index of intelligent forces "of which we know nothing except their existence as evidenced in these effects." Borrowing a page from Lyell, Argyll found a human analog for nature's hidden dynamic: "Languages grow, and change from generation to generation, according to rules of which the men who speak them are wholly unconscious. It is the same with all other things. . . . Even the work of Creation has been and is being carried on under rules of adherence to Typical Forms, and under limits of variation from them."[69] In short, "Typical Forms," such as the vertebrate archetype, made themselves manifest over time in the same way that the unknown but irresistible rules of phonetic change guided linguistic development.

In addition to archetypes, Argyll explained to British readers the other staple concepts of idealist morphology: homologies, "rudimentary" structures as anticipations of yet-to-be-introduced anatomical forms, and "correlation of growth." This last idea, denoting the phenomenon by which a variation in one part of an organism tended to give rise to a similar change in an anatomically related part, had been worked out by the Continent's leading transcendental anatomist, Etienne Geoffroy Saint-Hilaire (1772–1844).[70]

Finally, Argyll's most attractive evidence against the explanatory sufficiency of natural selection came from the brilliant coloring of hummingbirds. The colors marking its varieties played no apparent role in this species' struggle for existence. "Mere ornament and variety of form, and these for their own sake, is the only principle or rule with reference to which Creative Power seems to have worked in these wonderful and beautiful birds." This polemic greatly vexed Darwin, bringing home the need for a treatise on "sexual selection" to explain what natural selection could not.[71]

Darwin answered at least some of these arguments in his *Variation of Animals and Plants under Domestication* (1868). Here he reasoned that, if the minute variations that breeders cultivated could not be considered purposely designed, then neither could similar variations arising in untamed nature. Accordingly, in the book's concluding lines, Darwin explicitly rejected Asa Gray's viewpoint.

> No shadow of reason can be assigned for the belief that variations . . . which have been the groundwork through Natural Selection of the formation of the most perfectly adapted animals in the world, man included, were intentionally and specifically guided. However much we may wish it, we can hardly follow Professor Asa Gray in his belief "that variation has been led along certain beneficial lines," like a stream "along definite and useful lines of irrigation."[72]

Once again, however, Richard Owen came out with a new statement of his transcendentalist evolutionism, this appearing in the third (1868) volume of his *Anatomy of Vertebrates.* Owen declared variations to be teleological rather than random and held that the constant deviation of species from their ideal parental types took place according to a "defined and preordained course." Actually more skeptical than was his popularizer, Argyll, Owen refused to specify how variations were preserved, much less how they originated. In any case, Owen and Argyll set forth in the late 1860s the same essential reservations about Darwinian naturalism that Charles Lyell and Asa Gray had expressed in the early part of the decade. Each of these thinkers sought some kind of creational law behind evolution.[73]

An Analogic Zenith

The foregoing backdrop is needed in order to explain the presence and peculiar character of the linguistic analogies in *The Descent of Man.* Darwin completed the book's early chapters in the spring of 1870, only months after the reviews of August Schleicher's Darwinism pamphlet by Müller and Farrar had appeared in *Nature.*[74] Darwin's up-to-date reading would be apparent in the new work.

One of the linguistic images in *Descent* came purely as a result of Darwin repeating part of his argument from the *Origin,* that the principle of branching supplied a "natural" system of biological classification. He acknowledged that there were no records attesting the pedigree of any species, and that lines of descent could be discovered "only by observing the degrees of resemblance between the beings which are to be classed." The familiar analogy followed: "If two languages were found to resemble each other in a multitude of words and points of construction, they would be universally recognized as having sprung from a common source, notwithstanding that they differed greatly in some few words or points of construction."[75] Here Darwin quoted almost verbatim Sir William Jones's 1786 description of the idea that Greek, Latin, and Sanskrit had "sprung from some common source"—evidence perhaps that he had read Jones directly, although the phrase itself was in the air in that period.

Of greater importance were the analogies that appeared in *Descent*'s second chapter, "Comparison of the Mental Powers of Man and the Lower Animals." Here Darwin devoted nine pages to a subject more germane to his book, speculations about how speech might have originated. He summed up his viewpoint and his sources thus:

> With respect to the origin of articulate language, after having read on the one side the highly interesting works of Mr. Hensleigh Wedgwood, the Rev. F. Farrar, and Prof. Schleicher, and the celebrated lectures of Prof. Max Müller, on the other side, I cannot doubt that language owes its origin to the imitation and modification of various natural sounds, the voices of other animals, and man's own instinctive cries, aided by signs and gestures.[76]

It is interesting that Darwin included among these sources Schleicher's Darwinism pamphlet, even though Schleicher, neither in that work nor elsewhere, speculated on the actual origin of speech in the sense that the other three writers did. In any case, this reference suggests that Schleicher's analogy-laden pamphlet was on Darwin's mind. Darwin went on to describe a number of incipient language traits in animals and stressed how the gradual elaboration of these eventually could have led to human speech.[77]

What follows showed that evolutionism, conceived of as including humanity, was a doctrine ready-made for fudging categories. In this case, Darwin blurred reality into analogy by using the idea of gradual linguistic origins as a springboard toward a series of reflections on transmutation in general. He began with a sweeping declaration. "The formation of different languages and of distinct species, and the proofs that both have been developed through a gradual process, are curiously the same [revised edition: 'curiously parallel']." (Here Darwin cited Lyell's *Antiquity* for its "very interesting parallelism between the development of species and languages.")[78]

Even though his stated thesis involved the curious sameness of linguistic change and species evolution, Darwin turned immediately to the now familiar notion of an inequality between the two domains. "But we can trace the formation of many words further back than that of species, for we can perceive how they actually arose from the imitation of various sounds." On the surface, Darwin here continued to commend Wedgwood's and Farrar's views on the origin of language. Yet Darwin also borrowed from Schleicher the idea that words could be traced back to a much earlier stage of their developmental process than could the earliest formation of species. By this rhetorical sleight-of-hand, Darwin assumed as valid a naturalistic theory of language development while at the same time invoking Schleicher's notion that linguistic study had a leg up on paleobiology. The latter was a strange argument for a naturalist to make, yet it had the effect of suggesting that descent with modification was at least plausible.

Darwin then introduced completely new material, depicting anatomical patterns to which linguistic images had not yet been applied. He began with the signature concepts of Richard Owen's transcendentalist morphology. "We find in distinct languages (1) striking homologies due to community of descent, and (2) analogies due to a similar process of formation." On the linguistic side, as equivalents of biological homologies and analogies, Darwin presumably referred to the phenomena that students of language called by these same respective terms: cognate words, and similar paradigms of grammatical inflection, in related languages. If in languages, then also in organisms, Darwin suggested, these similar structures were "due to community of descent."[79]

Darwin also found that (3) "the manner in which certain letters or sounds change when others change is very like correlated growth." The first part of the reference here was to the sound correspondences found among cognate words in closely related tongues. As Jacob Grimm's law of phonetic shift demonstrated, initial *p* sounds in Latin words such as *pater* and *podos* nearly always correspond to *f* sounds in their Germanic cognates, as in *Vater* (father) and *Fuß* (foot). Darwin aptly compared such correspondences to variational symmetry in animal anatomy, what Geoffroy Saint-Hilaire had described as "the law of compensation or balancement of growth." Darwin found "still more remarkable" (4) the phenomena of rudiments "both in languages and in species." For example: "The letter *m* in the word *am*, means *I*; so that in the expression *I am*, a superfluous and useless rudiment has been retained. In the spelling also of words, letters often remain as the rudiments of ancient forms of pronunciation."[80] This of course expanded Darwin's analogy for vestigial organs appearing in the *Origin*.

Even by themselves, the four parallelisms described so far constituted a polemical tour de force, for in them Darwin attacked the entire transcendentalist interpretation of biological design. To Owen and Argyll, organic morphology revealed nature's intelligent "unity of plan." Yet Darwin explained these anatomical patterns—as the philologists did their counterparts in language—strictly in terms of common descent. The implication: no divine plan or creational force was needed.

Working upward from individual words to linguistic taxonomy, Darwin found that (5) languages, like organic beings, could be classed in groups under groups and that this could be done either artificially or naturally, that is, by descent. He thereby made a familiar distinction, assuming that a genealogical taxonomy of languages embodied a "natural" ordering not found in classification by grammatical typology.

Darwin then introduced an array of comparisons representing aspects of the transmutation process not yet treated in his writings. Several of these he borrowed from Lyell's *Antiquity* or Schleicher's pamphlet, and their biological analogs were fairly obvious. (6) The relationships between languages showed the effects of competition, as when "dominant languages and dialects spread widely, and lead to the gradual extinction of other tongues." (This idea also went back to Darwin's ethnological reading done in the 1850s.) (7) A language thereby made extinct constituted a dead end: like a species, it "never, as Sir Charles Lyell remarks, reappears. The same language never has two birthplaces." (8) Different languages could, however, be "crossed," that is, blended to form a distinct new tongue. (Darwin drew this theme from Farrar's "interesting article" in *Nature*.)[81] (9) Competition also acted upon variations arising within languages, because of new words "continually cropping up."

With this last point, Darwin finally alluded to the idea of natural selection; moreover, he vividly illustrated its well-known mechanism: "As there is a limit to the powers of the memory, single words, like whole languages, gradually become extinct." That is, proliferating words compete for the limited resource of human memory, just as population growth outstrips food supply in the Malthusian economy. Support for this parallelism came from a famous philologist. "As Max Müller has well remarked: 'A struggle for life is constantly going on among the words and grammatical forms in each language. The better, the shorter, the easier forms are constantly gaining the upper hand, and they owe their success to their own inherent virtue.'" Here Darwin culled what he found useful from Müller's review article in *Nature*.[82] Müller's description, he hinted, suggested that this "struggle" routinely eliminated a large number of lexical forms—individual organisms cast off from the mainstream of evolution. Müller had gone on to declare that the linguistic selection process entailed "rational elimination," yet Darwin found this part unconvincing and therefore omitted it. For as he well knew, thanks to Asa Gray, such wastefulness could not be considered a product of beneficent design.

The ninth point in Darwin's series of analogies described the most prevalent cause of the survival of some words rather than others. Yet one final contingency remained. (10) In some cases, words having a slight phonetic or semantic variation would be favored simply because they appealed to the mind's love of "novelty and fashion." This was Darwin's analogic answer to the duke of Argyll's hummingbirds, creatures that supposedly revealed nature's fondness for varied ornamentation as a good in itself. Surprisingly, Darwin did not equate point number 10 with sexual selection, although he may have intended to imply this; he would, after all, treat that topic in detail in part II of *Descent*. In any case, he capped off these final points with a magisterial declaration: "The survival or preservation of certain favored words in the struggle for existence is natural selection."[83]

With this single paragraph, Darwin gave free rein, as never before, to language-based analogies. Impressively, he broadened the image so as to cover the entire gamut of his theory from the overthrow of transcendentalist morphology to the support of his central explanatory device of selection. Of course, he had already communicated all of the natural-scientific points reflected here in a more direct fashion, in the *Origin*, in *Variation*, and in several parts of *Descent*. Thus really superfluous, *Descent*'s linguistic images nevertheless provided another opportunity for Darwin to inculcate these ideas through creative means.

In particular, *Descent*'s images contributed to the private tug-of-war among a select circle of naturalists and linguists. It had been spurred both by Darwin's enemies, Owen, Argyll, and Müller, and by his friends, Lyell, Gray, and Farrar. Darwin's relation to his friends in this regard was especially complex. Previously, Lyell had tried to counteract the *Origin*'s linguistic figures by using similar devices to promote a kind of argument from design. Darwin apparently welcomed the chance to recapture the image, inspired in part by Asa Gray's candid tutelage. Coached by Gray, he saw that the analogy not only suggested common descent but actually supported his antidesign view of nature: the innumerable trifling variations in the long history of language, so many of them eliminated rather than selected, were hard to reconcile with divine purpose.

Yet Darwin added a further twist via his double-edged use of the analogy, and in this he was probably inspired by Farrar. He obviously took material from Farrar's review in *Nature*, and he may also have read *Families of Speech*, which appeared at about the same time.[84] Imitating Lyell and Farrar, Darwin offered both linguistic parallels with transmutation and Darwinian explanations of the various facets of language development. In *Antiquity*, Lyell had introduced these theses one at a time by reversing the direction of the analogy in midchapter; Farrar, on the other hand, treated the two nearly simultaneously. Darwin followed the latter approach, even more so by making his comparisons equivocal from the very start. Once he declared that languages and species were "curiously parallel," each side of the equation served to explain the other. Hence the analogy gave the impression of internal sufficiency: in whichever direction it pointed, the comparison presupposed Darwinian evolution.[85] By its very structure, then, the *Descent* passage suggested that naturalistic mechanisms sufficed for producing both linguistic and biological development.

In this clever fashion, and with a flair for irony, Darwin once again turned the tables on Lyell, a repetition of the tactic he had used twelve years earlier in *The Origin of Species*. More than this, he essentially ratified Farrar's linguistic vision, which was based almost entirely on August Schleicher's ideas. The bond between the naturalist at Down House and the philologist at Jena, who never met or corresponded, had in this sense become a strong one. The larger lesson is simply that Darwin took such care with mere illustrative devices. With consistent attention to detail, he noted what his detractors used by way of imaginative figures and, whenever possible, turned these to his own purpose.[86] Again, he was especially quick to highlight any resonance, even in the use of analogies and metaphors, with Lyell's writings. Darwin still sought the great geologist's mantle even while undermining the contrary philosophy behind Lyell's natural science.

A final aspect of *Descent* deserving notice is the discussion of race in chapter 6, "The Affinities and Genealogy of Man." Here Darwin finally published thoughts that, years earlier, he had scribbled in the margins of the ethnological works he was reading.

> Now when naturalists observe a close agreement in numerous small details of habits, tastes, and dispositions between two or more domestic races [i.e., domestic animal or plant breeds], or between nearly-allied natural forms, they use this fact as an argument that they are descended from a common progenitor who was thus endowed; and consequently that all should be classed under the same species. The same argument may be applied with much force to the races of man.[87]

Mainly addressing the subject of human evolution, Darwin nevertheless made clear the link he perceived between the monogenesis debate and his general descent theory. On one side, he said, "those who do not admit the principle of evolution, must look at races as separate creations, or as in some manner distinct entities." (This ignored the fact that monogenist creationists had for nearly three millennia drawn the opposite conclusion.) On the other side, those naturalists who accepted the principle of evolution, "and this is now admitted by the majority of rising men, will feel no doubt that all the races of man are descended from a single primitive stock."[88] Darwin did not spell out his reasoning but assumed it to be self-evident that the logic of monogenesis went together with that of common descent.[89]

Although this thesis was not new to Darwin's thinking, the manner of its presentation was: *Descent*'s discussion of the race-species analogy appeared many pages distant from the linguistic analogies in the book's second chapter. The separation was intellectually fitting, for by then the revolution in human chronology had become well established. Linguistic descent and racial descent could no longer be conflated as they had been in the *Origin*'s ethnological analogy. This decoupling of themes reflected the larger separation in the 1860s between physical ethnology and the new cultural anthropology, with language now transferred to the latter sphere.[90]

The Question of Conscious Intent

The second half of the nineteenth century saw considerable debate as to whether the changes that languages undergo, many of them displaying clock-work regularity, were caused by human intention or by some overriding and mysterious force. We have encountered this question a number of times in passing: Charles Lyell's *Antiquity* chapter and the duke of Argyll's *Reign of Law* each referred to the unconscious character of patterned linguistic behavior; moreover, each linked this theme to supernatural design. As Argyll said, "Languages grow, and change from generation to generation, according to rules of which the men who speak them are wholly unconscious."[91] Although for professional rather than philosophical reasons, the philologists Schleicher and Müller said much the same thing, that linguistic change went its own way, unaffected by individual intent.[92]

There was no consensus on this issue, however, and even less agreement as to what it should imply for the Darwinian analogy. It all depended on which perspective best fitted one's views on evolution itself. In any case, this debate added an increased level of complexity and confusion to the analogic discourse on evolution; it also established further ties between the linguistic image and the larger structures of nineteenth-century social thought.

In the spring of 1870, just prior to the appearance of *The Descent of Man,*

British readers found this linguistic problem debated in *Nature*. Spurred by Max Müller's and F. W. Farrar's articles in that journal, one correspondent argued that the linguistic analogy was irrelevant: Darwinian transmutation entailed "*reasonless* variation and selection," whereas linguistic modifications were produced "by the countless efforts of *rational* beings." Another writer countered that the analogy still held good, "inasmuch as the 'gradual variation, etc., of a few primary sounds,' is not the result of an intention to originate a new language, any more than the origination of a new species of animal by natural selection is intentional on the part of the animals engaged in the struggle for life."[93]

In reply, the first correspondent admitted that there was no "grand intention" by individuals or societies to change their languages. Yet he reiterated that everyday language use involved "countless intentions of individual men to express individual ideas and thoughts." In light of this will to communicate in even the most reflexive speech act, one should "pause before entertaining a thought so revolutionary and perilous as that an eye, a beast, a man has been formed without presiding intelligence or design at all."[94]

The first writer, William Taylor of Stirling, summarized his case a year later in the *British and Foreign Evangelical Review*. He said that he had no objection "to the analogy pointed out by Professor Max Müller between the variation of living forms and the variation of languages." After all, there were plenty of well-known facts confirming that "languages *are* transformed and multiplied by a continual process of variation and survival."

> But here the analogy with thoroughgoing Darwinianism ends: for the impulses and efforts which have formed and which improve—we do not say merely alter— languages are not unconscious, but . . . are prompted and guided at every step by human reason. The analogy, as far as it applies, not only gives no support to the theory of the production of works of Intelligence by a process of unintelligent variation and selection, but is strongly against it.[95]

In short, although there was nothing wrong with applying Darwinian categories to language, the reverse—the linguistic image as representing Darwinism—was problematic theologically. Such were the vagaries of analogic discourse that the linguistic design argument had come full circle from the version found in Lyell's *Antiquity*. Lyell had stressed the unfathomable laws behind unintended linguistic change. Now an Evangelical spokesman, essentially on the same side as Lyell, argued the very opposite: Taylor emphasized the element of human rationality and volition in linguistic change, and for the purpose of the analogy, he conflated these things with divine superintendence.

The language-species comparison would reappear on occasion over the next half-century, sometimes invoked by leading scientists and scholars. Yet as an intentional argumentative tactic, it would no longer attract the degree of attention and debate that it did in the period from the *Origin* through *Descent*. In this sense, Darwin's 1871 book marked a kind of terminus. Paradoxically, however, the larger and unintended parallelism did not at all decline in importance. In a way, that story was only just beginning.

5

A Convergence of
"Scientific" Disciplines

THE LINGUISTIC IMAGES that grew thick in the debates over Darwinism from 1859 to 1871 appeared less frequently in the decades thereafter. Yet in a sense the analogy was stronger than ever, for, even when it went unspoken, it was nevertheless there. Indeed the comparisons examined in the preceding chapters were really but microcosms of a much larger parallelism. Those illustrations foreshadowed a convergence of concepts among the linguistic and biological sciences as a whole during much of the century after the appearance of Darwin's *Origin of Species*. What remains, then, is a narrative on a different scale: from now on, we will be interested less in individual analogies and more in the resonance between the larger disciplinary structures in which they became embedded.

A portent of this intellectual convergence issued from an unlikely locale, far from London, Jena, or Boston, and six years prior to the public advent of Darwinism. "Singular as it may now appear," said the South Carolina theologian James Warley Miles in 1853, "we do not hesitate to say, that when the genius of an Agassiz shall be happily united with the acquisitions of the scientific Philologist, the most striking analogies, and the most important conclusions will be derived from the combined deductions of Comparative Anatomy and Comparative Philology." The results of these two branches of science,

Miles predicted, "will, doubtless, ere long be brought into comparison and harmony."[1] This declaration recalls what we saw in chapter 1—the sense of a general parallelism between comparative philology and natural history even before 1859. The similarity between these fields would increase exponentially in the decades thereafter, although hardly in the way J. W. Miles envisaged. For the post-Darwinian convergence would manifest a distinctly genealogical pattern, interpreting change and taxonomy in terms of evolutionary common descent.

Recently, a number of writers have pointed out that the genealogical idea has structured both biological and linguistic knowledge. Yet they have tended to focus on the contemporary scene or else on its eighteenth- and early-nineteenth-century anticipations.[2] Actually, the similarity was most conspicuous in the post-*Origin* era, in the period from the 1860s until about 1930. On the one hand, these decades saw the construction of evolutionary ancestries come to dominate the practice of biological morphology. On the other hand, genealogy provided the conceptual foundation for that period's burst of growth in the linguistics-based cultural sciences: it served as a master metaphor in comparative mythology, comparative jurisprudence, "linguistic paleontology," and Indo-European studies in general.

It is a commonplace that this period also saw a more diffuse form of developmentalist thinking applied in the humanities and human sciences as a whole.[3] Yet this generic cultural evolutionism is not my concern here, and even less so the various social Darwinisms of that time. Neither do I attempt to survey the period's linguistic thought. Rather, I am interested only in cases in which a comparative method was used to reconstruct lost ancestral lineages. In all of this, I hope to demonstrate the impressive architectural similarity between a number of disparate fields of study. Equally important, I want to suggest the kind of perception of these fields' congruency that must have been created in the minds of informed lay people. My surmise, made more explicit in the epilogue that follows, is that this perception would have exerted a subtle yet profound influence on the scientific opinions held by lay readers. Essentially the language-species analogy writ large, it likewise would have reinforced the plausibility of biological transmutation.

The Schleicher-Haeckel Connection

Although I am concerned ultimately with large intellectual structures, my starting point is the personal interaction between two individuals, August

Schleicher and Ernst Haeckel, who molded that greater configuration to an immeasurable degree. Schleicher had been the first linguistic scholar to feature the *Stammbaum* or family tree diagram prominently, something he did on a small scale in two 1853 articles on the dispersion of the Indo-European *Urvolk*, and again in his 1860 treatise *Die Deutsche Sprache* (The Germanic Language). (see fig. 5.1). He then introduced his diagram representing all of the Indo-European tongues, from their beginning to the present (reproduced in chap. 4, above), in *Die Darwinsche Theorie und die Sprachwissenschaft* (1863). Moreover, Schleicher's two-volume *Compendium der vergleichenden Grammatik der indoger-manischen Sprachen* (1861–62), made the reconstruction of the Indo-European protolanguage a prime goal of historical linguistics.

These works, I argue, exerted a profound long-term influence not only on linguistics but also, surprisingly, on paleobiology. In other words, Schleicher's influence launched the trend of disciplinary convergence described in the present chapter. Schleicher of course actively sought to bring "Glottik" and biological science together to learn from each other and to render mutual aid: that was one of the purposes of his open letter to Ernst Haeckel, published as *Die Darwinsche Theorie*. Yet the century following the appearance of *The Origin of Species* also saw Schleicher's vision fulfilled in a way and with consequences beyond what he had intended. Still, the impelling event was the letter written to his scientific colleague at Jena.

Ernst Haeckel's fame in the history of science traditionally derives from his philosophical materialism and especially his emphasis on the "biogenetic law," the idea that ontogeny, the development of the individual embryo, recapitulates phylogeny, the evolutionary history of the parent species. For my purposes, however, Haeckel's importance derives not from this embryological doctrine but from his concept of phylogeny per se.

An ardent Darwinian and a man unencumbered by intellectual timidity, Haeckel invented the practice of reconstructing the ancestral lineages of zoological phyla. His first book, *Generelle Morphologie der Organisman* (1866), set forth what he called a *Descendenz-theorie*: it emphasized phylogenetic histories, these represented through family tree diagrams in the book's appendix. Similar diagrams appear throughout Haeckel's *Natürliche Schöpfungsgeschichte* (1868), a more popular work that went through several English editions as *The History of Creation* (1876). Some of these trees (all drawn by Haeckel himself) were vividly lifelike (see figs. 5.2, 5.3) while others were simple charts (see figs. 5.4, 5.5). In either case, they reconstructed the lineages of named taxa, linking living forms back to known yet extinct species and, prior to that, hypothesized ancestors. Haeckel, then, was the first true genealogist of living species.

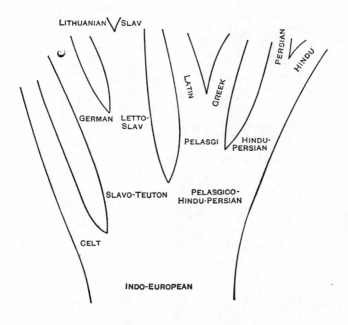

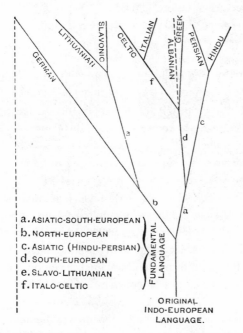

2.—PEDIGREE OF 1861 (1869). (*Die deutsche Sprache*², p. 82.)

Fig. 5.1. A. Schleicher, linguistic genealogies of 1853 and 1861. Reprinted in Otto Schrader, *Prehistoric Antiquities of the Aryan Peoples* (1890).

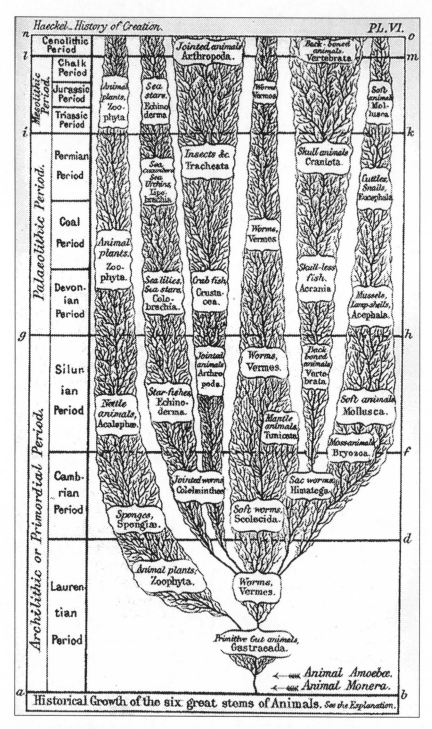

Fig. 5.2. E. Haeckel, six great stems of animals from *The History of Creation* (1868).

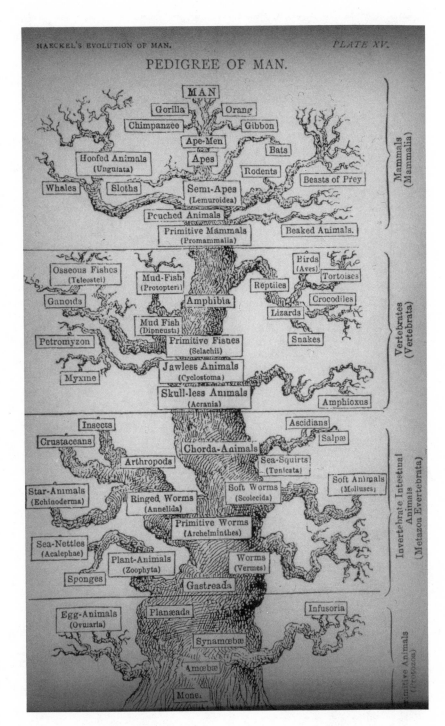

PEDIGREE OF MAN.

MAN

Gorilla   Orang

Chimpanzee   Gibbon

Ape-Men

Apes   Bats

Hoofed Animals
(Unguiata)

Rodents

Whales   Sloths   Beasts of Prey

Semi-Apes
(Lemuroidea)

Pouched Animals

Primitive Mammals
(Promammalia)   Beaked Animals.

Mammals
(Mammalia)

Osseous Fishes
(Teleostei)   Mud-Fish
(Protopteri)   Birds
(Aves)

Reptiles   Tortoises

Ganoids   Amphibia   Crocodiles

Lizards

Mud Fish
(Dipneusta)   Snakes

Petromyzon   Primitive Fishes
(Selachii)

Jawless Animals
(Cyclostoma)

Myxine   Skull-less Animals
(Acrania)   Amphioxus

Vertebrates
(Vertebrata)

Insects   Ascidians

Crustaceans   Salpæ

Chorda-Animals

Arthropods   Sea-Squirts
(Tunicata)

Soft Animals
(Molluscs)

Star-Animals
(Echinoderma)   Soft Worms
(Scolecida)

Ringed Worms
(Annelida)

Primitive Worms
(Archelminthes)

Sea-Nettles
(Acalephae)

Plant-Animals
(Zoophyta)   Worms
(Vermes)

Sponges

Gastreada

Invertebrate Intestinal
Animals
(Metazon Evertebrata)

Egg-Animals
(Ovularia)   Planæada   Infusoria

Synamœbæ

Amœbæ

Moner

Primitive Animals
(Protozoa)

Fig. 5.3. E. Haeckel, pedigree of man from *The Evolution of Man* (1874).

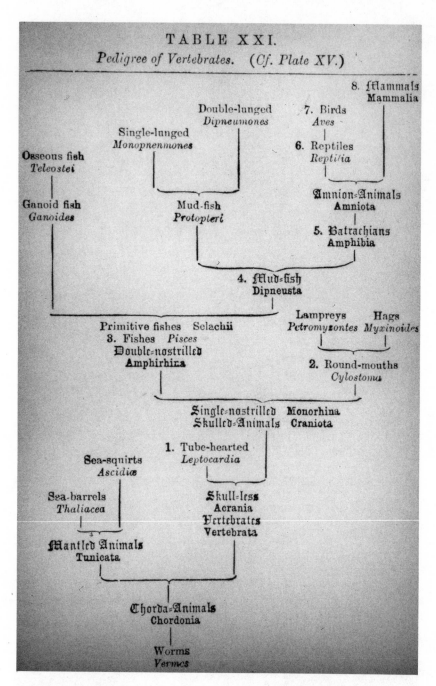

TABLE XXI.

Pedigree of Vertebrates. (*Cf. Plate XV.*)

8. Mammals
Mammalia

Double-lunged   7. Birds
Dipneumones   *Aves*

Single-lunged
Monopnenmones   6. Reptiles
Reptilia

Osseous fish
Teleostei

Ganoid fish   Amnion-Animals
Ganoides   Amniota

                    Mud-fish   5. Batrachians
                    Protopteri   Amphibia

                    4. Mud-fish
                    Dipneusta

Primitive fishes Selachii   Lampreys   Hags
3. Fishes *Pisces*   *Petromyzontes* *Myxinoides*
Double-nostrilled
Amphirhira
                                  2. Round-mouths
                                  Cyclostoma

Single-nostrilled Monorhina
Skulled-Animals   Craniota

1. Tube-hearted
Leptocardia

Sea-squirts
Ascidiæ

Sea-barrels   Skull-less
Thaliacea   Acrania
                   Vertebrates
                   Vertebrata

Mantled Animals
Tunicata

Chorda-Animals
Chordonia

Worms
Vermes

Fig. 5.4. E. Haeckel, pedigree of vertebrates from *The Evolution of Man* (1874).

TABLE XXIV.

Pedigree of Mammals.

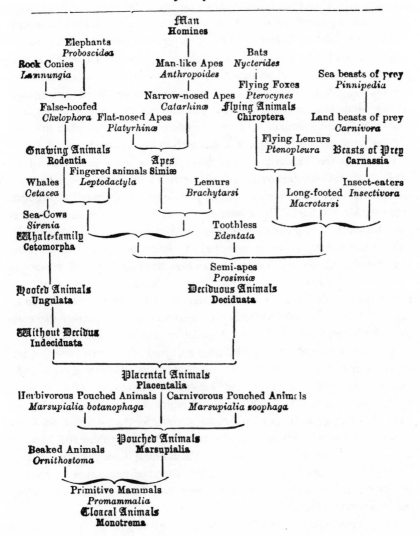

Fig. 5.5. Haeckel, pedigree of mammals from *The Evolution of Man* (1874).

Along with promoting the phylogeny concept, Haeckel gave considerable attention to the language-species analogy. He did this in *The History of Creation* and, more extensively, in *Anthropogenie* (1874), translated in 1903 as *The Evolution of Man*. In the latter work he devoted six pages to this "remarkable parallelism," hailing it as a "powerful ally" of the evolutionist cause. As far as he knew, the comparison had first been elaborated by August Schleicher, "not only a philologist but also a learned botanist." Haeckel commended Schleicher's *Darwinsche Theorie*, especially its linguistic pedigree diagram. Moreover, he included a similar diagram in his own book, placing it on a par with that work's organic pedigrees as well as its chart depicting the twelve "species" of mankind (see figs. 5.6, 5.7, 5.8). By juxtaposing these diagrams, Haeckel placed the language-species parallelism on display, bringing to an apex the practice of linguistic analogizing among naturalists.[4] Others would continue to articulate the comparison in writing, but no one else would etch its likeness as vividly.

Interest in the linguistic analogy, however, was but a small part of the intellectual debt Haeckel owed to Schleicher. In his 1874 book, Haeckel went so far as to embrace Schleicher's claim that philology held an investigative advantage over his own field: "The former can . . . adduce far more direct evidence than the latter, because the palaeontological materials of Philology, the ancient monuments of extinct tongues, have been far better preserved than the palaeontological materials of Comparative Zoology, the fossil bones of vertebrates. The more these analogous conditions are considered, the more convincing is their force."[5] This was a surprising admission for a naturalist to make, and it suggests the extent to which Haeckel accepted Schleicher's argument that comparative philology served as a kind of model for evolutionary natural history.

Yet there was more, for Haeckel even accepted Schleicher's view that language study had set the precedent for the reconstruction of biological genealogies: he declared that linguistics had "long ago anticipated in its own province the phylogenetic method with the aid of which we now attain the highest results in Zoology and in Botany."[6] Again, coming from a naturalist, this was a striking admission. It was not, I believe, merely a fulsome tribute to Schleicher, who had died in 1868 (though Haeckel was nothing if not fulsome). Rather, Haeckel was obliquely acknowledging that the chief inspiration for his own program of phylogenetic reconstruction was not primarily Darwinian biology but Schleicherian linguistics.

Most commentators on the Schleicher-Haeckel relationship focus on the importance of Schleicher's *Darwinsche Theorie* for Haeckel's philosophy of nature. Schleicher there expressed his belief in an ontological monism, the same outlook that would lace Haeckel's scientific writings.[7] In emphasizing Schleicher's influence on Haeckel's phylogenetic program, I follow the historian of linguistics Konrad Koerner. Koerner notes, first of all, that the circumstances of the two men's relationship suggest a kind of intellectual mentorship. Haeckel, thirteen years younger than Schleicher, came as a *Privatdozent* to Jena's Friedrich Schiller Universität, where Schleicher was an established professor. Schleicher was then at the height of his international reputation, having just published his *Compendium*. Within a few years, Haeckel would refer to Schleicher as among the closest of his Jena friends.[8]

Perhaps Haeckel was especially impressed by his colleague's intellectual independence. Prior to reading *The Origin of Species*, Schleicher had already familiarized himself with botanical science and had set forth his biological theory of language. As he noted in *Die Darwinsche Theorie,* he had acquired his perspective from reading J. M. Schleiden's *Botanik als induktive Wissenschaft* (1843) and Carl Vogt's *Physiologischen Briefe* (1845–47). In other words, Schleicher declared that Darwin merely confirmed what he already knew. To him, Darwin's theory was not so much a work of individual genius as something called forth by the spirit of the age, a "legitimate child," as he said, of the nineteenth century. All of this must have given Haeckel a strong sense of Schleicher's authority and scientific credentials.[9]

Yet can Schleicher's impact on Haeckel's phylogenetic imagination be pinned down? Apart from the obvious similarity of their pedigree concepts, the case for this historical linkage revolves around timing. Haeckel's *Generelle Morphologie* appeared in 1866, only three years after Schleicher's open letter, and there are no other likely candidates for direct influence in this period. Another Jena colleague, the zoologist Carl Gegenbaur, also became a formulator of evolutionary pedigrees. Yet he began publishing work in this line only in 1870, well after Haeckel had led the way.[10] Some have suggested that Haeckel's program was simply a natural outgrowth of Darwin's theory, even though Darwin himself never attempted anything like this in his published writings.[11] Yet this view only highlights the question of why Haeckel, apart from his temperamental enthusiasm, became the first to take the leap. Schleicher therefore appears as the most likely source of intellectual encouragement for Haeckel's program. In this light, the historical significance of their friendship is enormous.

That personal connection, given intellectual expression in Schleicher's essay on Darwinism and linguistics, affected not only the future direction of paleontological research, this being the most obvious result, but also formed the wellspring of the disciplinary convergence described in this chapter. There was in this sense nothing fortuitous about the post-*Origin* congruence between philology and biology. At its source, it was a matter of the most direct kind of

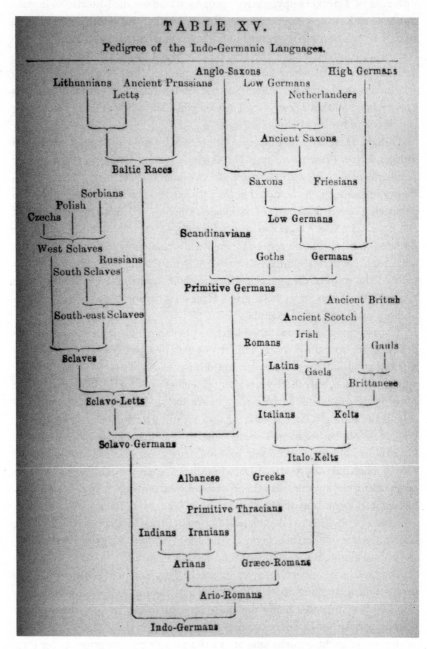

Fig. 5.6. E. Haeckel, pedigree of the Indo-Germanic languages from *The Evolution of Man* (1874).

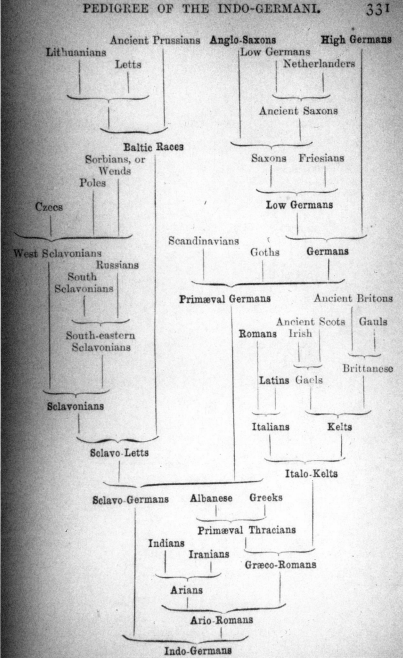

Fig. 5.7. E. Haeckel, pedigree of the "Indo-Germanic" peoples from *The History of Creation* (1868).

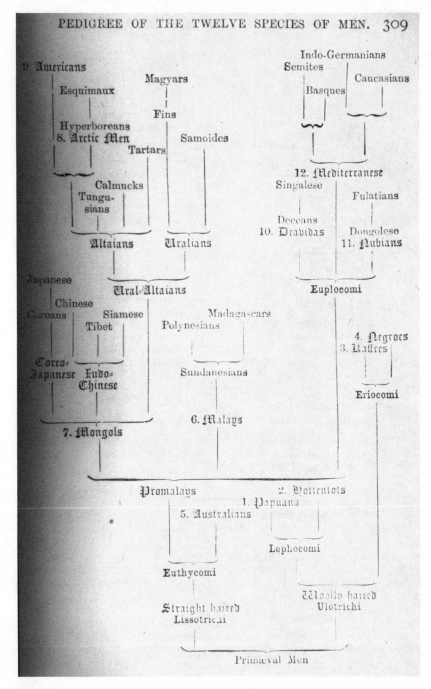

Fig. 5.8. E. Haeckel, pedigree of the "Twelve Species of Man" from *The History of Creation* (1868).

cross-disciplinary influence. It should be noted, however, that once the practice of constructing family trees gained a foothold in biological science, its link back to Schleicher and comparative philology for the most part became obscured.

Continuations of the Analogy

Before proceeding with the examination of the unspoken congruence between biology and philology, we should note how a number of writers, in addition to Ernst Haeckel, continued to use the explicit language-species analogy in the late nineteenth and early twentieth century. Naturalists now used the linguistic image to represent common descent alone, a falling away from the high point of analogic complexity found in Darwin's *Descent of Man* (1871), with its parallels to natural selection and morphological patterns. This new restriction of the analogy was apt: it mirrored the widespread acceptance of the branching descent pattern despite disagreements over other aspects of Darwin's theory.

Apart from Haeckel, the other leading naturalist to draw the analogy was George John Romanes (1848–94), Darwin's last face-to-face disciple. Romanes reintroduced the comparison in 1892, yet with two significant twists. First, Huxley-like, he loaded it with antitheistic import—although this did not stop him from invoking the memory of Lyell: "Classification of organic forms, as Darwin, Lyell, and Häckel have pointed out, strongly resembles the classification of languages. . . . Now what would be thought of a philologist who should maintain that English, French, Spanish, and Italian were all specially created languages—or languages separately constructed by the Deity . . . and that their resemblance to the fossil form, Latin, must be attributed to special design?"[12]

Romanes's second new theme came despite his show of deference for Haeckel.[13] Writing from the perspective of the 1890s rather than the 1860s, he overturned the advantage thesis that the German zoologist had inherited from Schleicher. Even though philology well illustrated the descent idea, "the evidence of the natural transmutation of species is in one respect much stronger than that of the natural transmutation of languages—in respect, namely, of there being a vastly greater number of cases all bearing testimony to the fact of genetic relationship." Here Romanes portrayed his own field, not linguistics, as the evolutionary success story, one providing a wealth of testimony about genealogical relationship. Yet this declaration amounted to a last hurrah, for use of the analogy among naturalists waned after this time. The only other example I have found appeared in a 1917 book by the Princeton University paleontologist William B. Scott.[14]

Yet there was also the opposite version of the analogy, the Darwinian model applied to language as seen in Lyell's *Antiquity* chapter, in Schleicher's *Darwinsche Theorie,* and in Darwin's *Descent* paragraph. This side of the analogy did its part to heighten the sense of an overall reciprocity between the two domains. The notion of Darwinian patterns in language became common coin in the late nineteenth century: as F. W. Farrar said, the idea suggested itself so naturally as to make its use almost inevitable. The American philosopher and Darwinist Chauncey Wright accordingly noted the "precise parallels" between linguistic and organic evolution.[15] One of Wright's colleagues in the Cambridge Metaphysical Club, the scientific writer and sometime Harvard lecturer John Fiske, also mentioned the analogy. Like Farrar, however, Fiske elevated it to the level of the entire disciplines involved.

> It was by the establishment of genera and species that the study of words, as well as the study of organized beings, first assumed a scientific character. Comparison in the hands of Bopp played as conspicuous a part as in the hands of Cuvier, and the notable resemblance between the results obtained in philology and those recently arrived at by Darwin has been frequently remarked and sometimes unduly insisted upon. The two sciences, indeed, utterly diverse as are their subjects of research, are wholly alike in their methods, and the science of language will do well not to neglect the useful hints which she may often receive from the experience of her older sister.[16]

In short, two utterly different sciences were alike in both method and results. Although Fiske kept biology in the dominant position, he was portraying a conceptual congruence between two entire fields of inquiry.

Especially authoritative was the use of the analogy by the linguistic experts themselves. Among the major philological writers, Max Müller, the American William Dwight Whitney, and the Frenchman Abel Hovelacque all remarked on the parallel with Darwinism—although usually adding a warning that it should not be taken too literally.[17] Some did question the analogy, at least implicitly. In 1875, Oxford's Archibald Henry Sayce (1845–1933) followed Max Müller in denying the reality of unified parent languages from which daughter tongues had diversified. Sayce also denied the notion of linguistic evolution through the isolating, agglutinating, and inflective types of morphology.[18]

The French linguist Abel Hovelacque fought back the next year with an explicit defense of the Darwinian parallel: he was certain that the agglutinating languages had been isolating in structure prior to reaching their present condition and that the inflecting languages had "successively passed through the two previous stages." The objection to this conclusion had arisen from the absence of intermediate links between these known grammatical types—a familiar kind of argument. "We have not here to pronounce on a question of zoology or botany," Hovelacque responded, "but we would remark that where language is concerned the objection has no force whatsoever, for the process of evolution is here easily followed, and in fact detected in active operation. The transmutation of species is here a patent fact, and one of the fundamental principles of the science of language."[19] That is, even if evolution were not true in biology, at least it was so in philology.

The most prominent Continental philologist to commend this version of the analogy was Hermann Paul, a leader of the influential Neogrammarian school of historical linguistics. In *Prinzipien der Sprachgeschichte* (1880), translated in England as *Principles of the History of Language* (1888), Paul did battle with August Schleicher's contention that languages, like natural organisms, underwent change apart from the influence of individual agency. For his part, Paul drew a Darwinian picture of language that suggested the very opposite: his central concept was what linguists now call idiolects, the unique speech pattern of each individual. This differentiation at the level of the individual, Paul argued, provided the ultimate source of that diversifying impulse that led relatively homogeneous languages to branch into dialects and then into whole new tongues. It was the same in Darwinism, which saw large-scale changes arising ultimately from individual variation.[20]

What is important here is not so much the philologists' ongoing debate concerning conscious agency. Rather, it is how, with a bit of ingenuity, a Darwinian model of change could be pressed into service on either side of that debate. Either way, linguistic writers tended to accept some form of the analogy, thereby reinforcing the general logic of common descent.

A few other instances of this kind appeared in the early twentieth century. The Cambridge University classicist J. M. Edmonds compared the process of reconstructing a "hypothetic language" to that of reconstructing the parent species of the horse. The theme also received honorable mention from Edmonds's colleague Peter Giles, in his contribution to a Darwin centennial volume in 1908.[21] With this, the vogue of the Darwinian analogy in linguistics had passed, in a way similar to that in which the reverse parallelism faded in biological writings. Yet this waning of the actual comparison was more than offset by the growing congruence between the encompassing disciplines.

The Parallel Paths of Biology and Linguistics

Ernst Haeckel's program of phylogenetic reconstruction quickly inspired a following in England. T. H. Huxley was won over almost immediately, even though the practice was taboo for Darwin: the well-known tree diagram in the *Origin of Species* presented only the most abstract idea of a paleobiological pedigree. After looking over the German edition of Haeckel's *Natürliche Schöpfungsgeschichte*, Darwin told its author: "Your chapters on the affinities and genealogy of the animal kingdom strike me as admirable and full of original thought. Your boldness, however, sometimes makes me tremble, but as Huxley remarked, some one must be bold enough to make a beginning in drawing up tables of descent."[22]

Huxley's words to this effect were accompanied by deeds of his own, although he tried to protect himself against the charge of excessive speculation. Already, in an 1867 paper, he had displayed the affinities among bird species according to what he called "classification by gradation." Huxley admitted that this arrangement was not truly genetic, for it did not actually express "the manner in which living beings have been evolved one from the other." Yet he still held the latter to be "the ultimate goal" of paleontological zoology, and he suggested that his own schema was an approximation of this. He also hailed the appearance of Haeckel's 1868 book. Although guarded in his praise, Huxley stressed the volume's "most original" section, the one that worked out the general evolutionary lineage of the two organic kingdoms. Here, he said, the German naturalist had furnished each kingdom with "its proper genealogical tree, or 'phylum.' "[23]

Other researchers joined in following Haeckel's program: during the 1870s and 1880s especially, taxonomists from Vienna to Boston rushed to propose evolutionary genealogies of particular zoological groups. This genealogical emphasis was emblematic of the curious history of Darwinism in the half-century after publication of the *Origin of Species*. In that era, common descent became common ground for naturalists who disagreed as to the mechanism and direction, if any, of evolution; hence Darwin's descent thesis prevailed even as natural selection took a back seat as an explanatory mechanism.[24] The *Origin* had meshed these factors in an integrated argument, especially in the chapter on natural selection itself: significantly, it was here that Darwin chose to introduce his tree-of-descent diagram and to discuss the phenomenon of divergence of character.[25] Yet other naturalists would separate these factors, retaining some and leaving others.

Darwin himself set the precedent for this by subdividing his theory and ranking its components in order of importance. He did this, for instance, when he complained in private about Lyell's equivocal reaction to his theory in *Antiquity of Man*: "I have sometimes almost wished that Lyell had pronounced against me. When I say 'me,' I mean only change of species by descent. That seems to me the turning point. Personally, of course, I care much about Natural Selection: but that seems to me utterly unimportant compared to [the] question of *Creation or Modification.*" Darwin was only repeating what he had just said in public. In an 1863 letter to the London *Athenaeum*, he had suggested that the pursuit of an evolution-based research program did not depend upon the acceptance of his entire theory: "Whether the naturalist believes in the views given by Lamarck, by Geoffroy St.-Hilaire, by the author of the 'Vestiges,' by Mr. Wallace and myself, or in any other such view, signifies extremely little in comparison with the admission that species have descended from other species and have not been created immutable; for he who admits this as a great truth has a wide field opened to him for further inquiry."[26]

Of course, Darwin hoped that his own views would be followed more closely than this, and he was greatly disappointed when the division of his theory's components and the relative demotion of natural selection became the actual practice of many of his followers. Most prominently, Huxley had his doubts. Still, these researchers kept closer to the Darwinian pattern than even Darwin himself had suggested, for they did not simply accept modification as such but fastened in particular upon common descent. The idea of branching evolution thus came to guide the research programs of comparative anatomy, comparative embryology, systematics, and biogeography, fields that supplied the bulk of the evidence pointing toward transmutation for several decades after the *Origin*.[27]

This consensus around the notion of biological genealogy did not amount to a unanimity of opinion on all particulars.[28] Haeckel's phylogenies were often judged to be too conjectural, and naturalists felt chastened by the end of the 1880s because of the lack of agreement in their reconstructions. Yet workers in that field still aspired eventually to produce full and reliable evolutionary pedigrees. Efforts continued especially to place fossil data into genealogical schemata: the ancestry of the horse was late-nineteenth-century paleontology's greatest success story of this kind. Moreover, this emphasis bridged the divide produced by the rise of genetics-based neo-Darwinism in the early twentieth century.

The continuing focus on common descent was put on display via pictorial phylogenies, these appearing in books written by a roll call of leading British and North American naturalists: Edward Drinker Cope (1887), Edwin Ray Lankester (1910), Arthur Dendy (1912), Henry Fairfield Osborn (1918), J. B. S. Haldane and Julian Huxley (1927), George Gaylord Simpson (1944, 1953), and William King Gregory (1951).[29] Descent diagrams were also exhibited in works aimed at a more popular audience.[30] As the above list shows, the genealogical

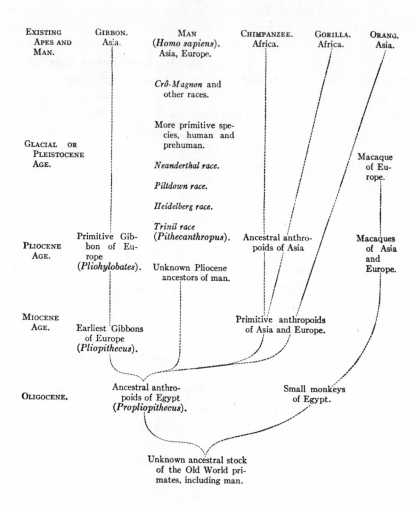

Fig. 5.9. H. F. Osborn, ancestral tree of the anthropoid apes and of man from *Men of the Old Stone Age* (New York, 1915).

concept held constant throughout the post-*Origin* century (and until today as well), forming the iconic image of biological evolutionism in the public mind.[31]

That image appeared in a colorful variety of forms, all of them reinforcing the unity of the basic pattern (see figs. 5.9, 5.10). In New York City, for instance, visitors to the American Museum of Natural History's Hall of the Age of Man would for many years see displayed a chart depicting the genealogical relation-

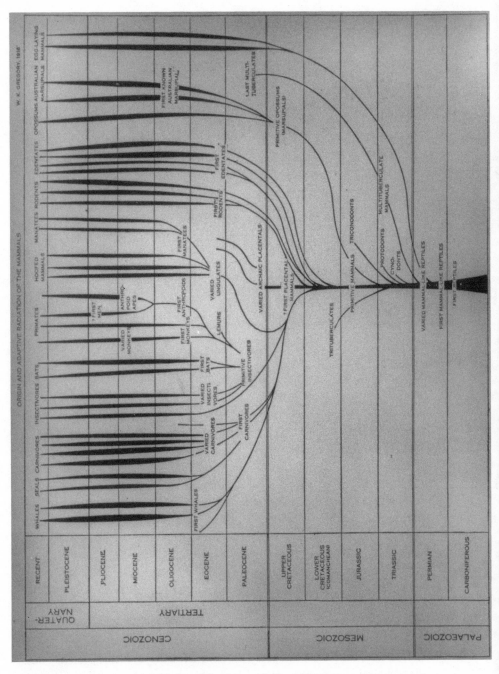

Fig. 5.10. H. F. Osborn, ancestral tree of the mammals from *The Origin and Evolution of Life* (1917).

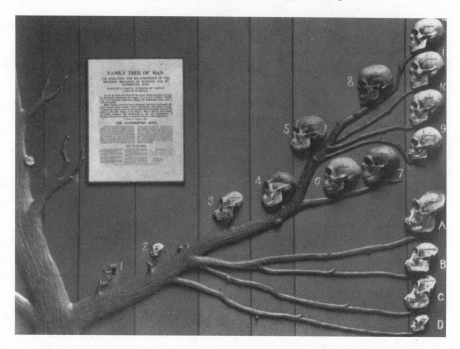

Fig. 5.11. H. F. Osborn, the family tree of man (1927). Image no. 310635, courtesy Department of Library Services, American Museum of Natural History.

ship among anthropoid skull types (see fig. 5.11). The family tree concept received perhaps its most explicit endorsement in the frontispiece to a popular 1928 exposition of evolutionary thought. This volume gave only slight attention to the concept of natural selection yet placed on pictorial display "The Tree of Life" (see fig. 5.12). These visual pedigrees must have produced an impact comparable to that made by the frontispiece of Huxley's *Man's Place in Nature* (1863) picturing the skeletal profiles of a gibbon, orang, chimpanzee, gorilla, and human. Both kinds of image showed at a glance what the experts agreed to be the historical pattern of biological change.

Both in research methodology and in its popular iconography, the field of linguistics employed genealogical assumptions similar to those found in the life sciences. This was true despite the ironic path that the genealogy concept traveled in linguistics. August Schleicher's celebrated *Stammbaum* theory of Indo-European development came under an attack, from which it never really recovered, a little over a decade after Darwin's *Origin of Species* appeared. (I note the technical arguments involved later in this chapter.) Nevertheless, the basic

CREATION BY
EVOLUTION

A CONSENSUS

OF PRESENT-DAY KNOWLEDGE AS SET FORTH
BY LEADING AUTHORITIES IN NON-TECHNI-
CAL LANGUAGE THAT ALL MAY UNDERSTAND

EDITED BY

FRANCES MASON

NEW YORK
THE MACMILLAN COMPANY
MCMXXVIII

Fig. 5.12. "The Tree of Life," frontispiece from F. Mason, ed., *Creation by Evolution* (1928).

comparative-genealogical method that the *Stammbaum* represented, as well as family tree diagrams, continued to be used on through the twentieth century.

Linguistic tree diagrams, the popular face of linguistic genealogy, appeared often in works of general reference. Many nineteenth-century dictionaries, for example, reprinted Samuel Johnson's 1755 historical sketch of the English language, including its chart showing the derivation of English from "Gothick or Teutonick." Although an attested language, Gothic was only barely so. Dr. Johnson thereby set a significant precedent: he publicized the notion of descent from a remote linguistic ancestor that only the experts could conjure up.

By the beginning of the twentieth century, dictionaries regularly exhibited pedigrees of the entire Indo-European family, showing its taproot in the "Aryan Mother Tongue." Now readers were asked to rely even more on the philologists' competence, in this case with regard to a past which was purely conjectural. Simple in construction, these pedigree charts nevertheless carried all of the scholarly authority of the volumes that displayed them, and they must have imprinted the scientific imagination of even the most casual reader. This was not a matter of any overt reference to Darwinian biology; the discussion surrounding these diagrams almost never mentioned the parallelism. Still, the visual similarity would have spoken for itself. The likely—although perhaps only semiconscious—impact of the language pedigrees may be gauged by comparing the reproductions in figures 5.13, 5.14, and 5.15 with the biological charts shown in figures 5.9, 5.10, and 5.11.

Philological practice undergirded these pictorial representations. Most important here was the elaboration of the Indo-European idea. Beginning with Schleicher's work in the 1850s and 1860s, comparative philology took a quantum leap in technical sophistication. August Fick's *Wörterbuch der indogermanischen Grundsprache* (1868) continued Schleicher's program by setting up principles for reconstructing proto-Germanic. Even if that tongue proved to be a philologists' fiction, this did not invalidate the reconstructive project. This was true especially of the idea of historical-linguistic laws.

The 1870s saw the rise of an influential new school of German comparativist researchers, the so-called *Junggrammatiker* or Neogrammarians.[32] These linguists saw themselves as rebels, yet they built extensively on the work of their predecessors: they refined Jacob Grimm's analysis of Germanic sound correspondences in order to formulate exact new linguistic laws. All such correspondences emerged over time as phonetic shifts, uniform changes in the articulation of a group of similar words which bore witness to the common descent of the affected languages. These laws became the keys to establishing both kinship and classification: the degree of relatedness between a given set of languages was directly proportional to the number of correspondences they consistently shared.

Sound laws also became the primary means by which protoreconstruction was carried out: they allowed philologists to identify rule-bound changes, conjecturally reverse those change processes, and thereby reconstruct the phonological system of an ancestor tongue for which there was no direct evidence.[33] Hence the Swiss linguist Ferdinand de Saussure (1857–1913), known later for the structuralist themes in his *Cours de Linguistique Générale* (1916), achieved fame in his own lifetime for a comparative-historical treatise: *Mémoire sur le système primitif des voyelles dans les langues indo-européennes* (1879). Henry Sweet, England's leading nineteenth-century phonetician, described this reconstruction of the proto-Indo-European vowel system as "perhaps, the most important event that has happened in the history of comparative philology since its foundation."[34]

The comparative method could be put to further uses once a set of protoforms had been reconstructed. In Romance studies, Indo-European protoforms proved useful in recovering the vernacular dialects of Latin, a language attested mainly in its literary form. (Thus confirming the healthy kernel within Max Müller's dialects thesis.) This procedure has been described as "reconstructing forward": starting with the known features of a conjectured prototongue, philologists reconstruct the changes that produced the less-well-understood daughter languages.[35]

A remarkable application of this technique appeared in biblical archaeology. As director of the American School of Oriental Research during the 1920s, William F. Albright (1891–1971) engaged in historical "site identification"—using existing Palestinian place-names as keys to finding long-lost biblical sites. Albright dismissed the approach to this task taken by its pioneers, who had relied on the most superficial name resemblances. "The early investigators, as well as all too many recent ones, who never seem to have heard of philological law . . . , allowed themselves unlimited license."[36] The solution lay in the rules of Semitic sound change, rooted in the notion of a Semitic family of languages in which Hebrew and Arabic were siblings with a common ancestor. Phonological laws would show whether an Arabic place-name actually preserved some vestige of an ancient Hebrew name recorded in the Bible. Could the researcher predict that the older Hebrew sound would have evolved into the sounds appearing in the later Arabic name? If he could not, then even the closest similarity between the names was merely fortuitous, and archaeologists had no right to identify on linguistic grounds the present-day site with the proposed biblical candidate. In essence, then, the philologist inferred what could or could

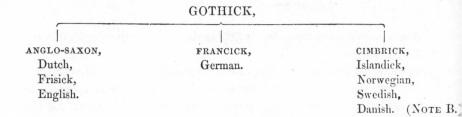

GOTHICK,

ANGLO-SAXON,
Dutch,
Frisick,
English.

FRANCICK,
German.

CIMBRICK,
Islandick,
Norwegian,
Swedish,
Danish. (NOTE B.)

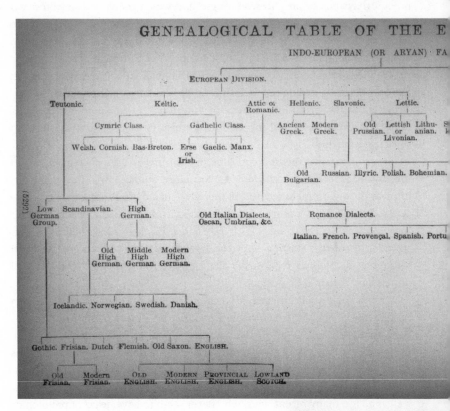

GENEALOGICAL TABLE OF THE E

INDO-EUROPEAN (OR ARYAN) · FA

EUROPEAN DIVISION.

Teutonic.   Keltic.   Attic or   Hellenic.   Slavonic.   Lettic.
Romanic.

Cymric Class.   Gadhelic Class.   Ancient   Modern   Old   Lettish   Lithu- S
Greek.   Greek.   Prussian.   or   anian.   L
Livonian.

Welsh. Cornish. Bas-Breton.   Erse   Gaelic. Manx.
or
Irish.

Old   Russian. Illyric. Polish. Bohemian.
Bulgarian.

(§329)   Low   Scandinavian.   High   Old Italian Dialects,   Romance Dialects.
German   German.   Oscan, Umbrian, &c.
Group.

Italian. French. Provençal. Spanish. Portu

Old   Middle   Modern
High   High   High
German. German. German.

Icelandic. Norwegian. Swedish. Danish.

Gothic. Frisian. Dutch Flemish. Old Saxon. ENGLISH.

Old   Modern   OLD   MODERN   PROVINCIAL   LOWLAND
Frisian.   Frisian.   ENGLISH.   ENGLISH.   ENGLISH.   SCOTCH.

Fig. 5.15. genealogy of English,
from H. C. Wyld, *The Universal
Dictionary of the English Language*
(1932).

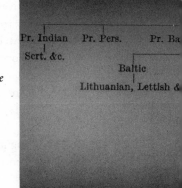

Pr. Indian   Pr. Pers.   Pr. Ba
Scrt. &c.
Baltic
Lithuanian, Lettish &

Fig. 5.13. S. Johnson, genealogy of English, reprinted in R. G. Latham, *A Dictionary of the English Language* (1876).

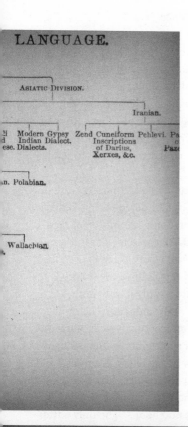

LANGUAGE.

ASIATIC DIVISION.

Iranian.

Modern Gypsy Zend Cuneiform Pehlevi. Pa
Indian Dialect.   Inscriptions
ese. Dialects.   of Darius,   Paz
           Xerxes, &c.

n. Polabian.

Wallachian

Fig. 5.14. genealogical table of the English language, from R. Hunter, *The Encyclopedic Dictionary* (1896).

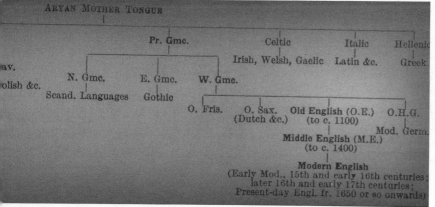

ARYAN MOTHER TONGUE

Pr. Gmc.   Celtic   Italic   Hellenic

               Irish, Welsh, Gaelic Latin &c.   Greek

av.   N. Gmc.   E. Gmc. W. Gmc.
olish &c. Scand. Languages Gothic

                  O. Fris.   O. Sax. Old English (O.E.) O.H.G.
                     (Dutch &c.) (to c. 1100)
                                         Mod. Germ.
                         Middle English (M.E.)
                          (to c. 1400)

                          Modern English
              (Early Mod., 15th and early 16th centuries;
                later 16th and early 17th centuries;
            Present-day Engl. fr. 1650 or so onwards)

not have happened in the historical blank space between data from two different eras. In cases where linguistic laws came into play, one could reach hypothetico-deductive conclusions concerning empirically unattested events.

Patterns of Tacit Congruence

We are now in a position to explore more deeply the methodology and pattern of results that bound together Darwinism and comparative-historical linguistics. No single theoretical statement or practical example covers the entire logical tapestry involved in this disciplinary convergence, for it was woven from a number of originally separate patches. Here, then, I take the liberty of ranging freely over times and texts so as to tighten the circle of relevant ideas. By juxtaposing those themes that seem most closely connected, I hope to render intelligible this cross-disciplinary schema.

I begin with the most abstract aspect of this shared intellectual structure. In writing to the London *Athenaeum* in 1863, Darwin had defended his descent theory by stressing its ability to account reasonably for a "multitude of facts"; he touted as well its utility in stimulating further research. These themes harked back to his transmutation notebooks of 1838, where Darwin described his method: "The line of argument often pursued throughout my theory is to establish a point as a probability by induction and to apply it as hypotheses to other points and see whether it will solve them."[37] An induction, he implied, could be generalized only by means of further inductions; yet because this process was never exhaustive, it could never attain certainty in the deductive sense—the proof that follows necessarily from a given set of premises—and therefore yielded only probabilities. Well-tested probabilities, however, could themselves serve as hypotheses to guide further inquiry. This reasoning applied even to scientific laws: as Darwin remarked in his notebook, "The only advantage of discovering laws is to foretell what will happen and see [the] bearing on scattered facts." *Law* in this sense allowed the work of science to go forward by the prediction and subsequent testing of research findings.[38] Did the study of language offer any parallel to this kind of thinking?

A. H. Sayce, Max Müller's protégé at Oxford, considered the matter of scientific method in the introduction to his *Principles of Comparative Philology* (1875). Sayce made no reference to Darwin, yet his comments bore a striking resemblance to those just quoted. He described his own field as a "strictly inductive" historical science, not an "exact deductive science" like astronomy. (Here Sayce distanced himself from Müller's view that comparative philology constituted a natural science.) Like all scientific disciplines, philology aspired to produce laws of operation; still, philology being a historical field, these so-called laws could never be made absolutely precise. Laws and hypotheses, moreover, were valuable mainly for stimulating additional inquiry: they "supply the mind with a clue for further researches; they serve to connect the isolated facts, and to simplify the bewildering maze in which we find ourselves."[39] As Sayce implied, this kind of hypothesizing was closely tied to philology's retrospective, historical orientation.

On the biological side, one of Darwin's early defenders made a similar conceptual link between the formation of testable hypotheses and the reconstruction of the past. The argument appeared in the opening pages of Fritz Müller's *Für Darwin* (1864), translated as *Facts and Arguments for Darwin* (1869). In order to verify Darwin's views, Müller said, researchers needed to apply them to the developmental history of some particular group of organisms, such as crustacea. They needed, that is, to try to establish a "genealogical tree" and "to produce pictures as complete and intelligible as possible of the common ancestors of the various smaller and larger circles" of that group. This effort would produce one of three results: it would show irreconcilable and contradictory outcomes of Darwin's theory, it simply might not work at all, or it would successfully demonstrate descent.[40]

These few remarks by Darwin, Fritz Müller, and the philologist Sayce, show the essential contours of this shared mode of hypothesizing. First, all parties assumed that conjectures about an irrecoverable past could nonetheless be scientifically valid. Moreover, those past-oriented conjectures produced results that inevitably took on a genealogical pattern.[41]

The *Stammbaum* metaphor was appropriate here since the passage from trunk to twigs indicated diversity emerging out of unity: an ancestor could have many descendants whereas a given descendant could have but one immediately preceding ancestor. The genealogical schema therefore ruled out convergent development or hybridization.[42] It also had a double-edged character: it entailed a retrospective mode of inference that was largely conjectural yet could claim more certainty than did outright speculation. That is, it compensated for its inexactness by channeling inquiry into pre-set grooves, greatly simplifying the task of historical reconstruction.[43] This approach to the past confined itself to the tightly structured realm of genealogy as opposed to the contingent and capricious realm of real history. It was a controlled mode of retrospection that had little to do with the historical discipline proper, even including the "philosophical" historiography admired by many nineteenth-century European intellectuals.

Still another overlap between Darwinism and philology was created by their shared use of a comparative method, a triangulation between more or less contemporaneous phenomena in order to extrapolate toward some earlier condition. This procedure implied certain prior assumptions about the developmentalist origins of present diversity. It is helpful in this connection to regard the comparative method as a species of conjectural history. The historian F. J. Teggart perceptively characterized this practice: it envisaged "the differences with which we are confronted in the present world as the product of changes which have taken place in the past." The procedural technicalities of the method are less important than its end product: a reconstruction of temporal antecedents which becomes the basis for classifying present-day diversity.[44]

Recent biological systematics suggests further insight into the logic shared by nineteenth-century linguistics and Darwinism. Beginning around the middle of the twentieth century, the "cladistic" school breathed new life into the idea of classification based on evolutionary pedigrees. Significantly, one of this school's leaders describes a method of discovery similar to that used in both paleobiology and philology in the late nineteenth century.[45] As a graduate student in the 1940s, the University of Michigan botanist Warren H. Wagner studied the genus *Diellia*, a small set of Hawaiian ferns. He formulated a "cladogram" (a phylogenetic diagram) showing the degrees of kinship among the species of this group, and, on this basis, described a generalized hypothetical ancestor. Wagner assumed that this reconstructed ancestor corresponded to some extinct species, beyond empirical recovery.

> However, during a post-doctoral year at Harvard following the completion of my thesis, while working over unidentified species at the Gray Herbarium, I discovered a fern that conformed to my predicted ancestor. To be true, the specimen was only a parched herbarium mummy of a collection made in 1879 in Makawao, Maui. This plant had been confused with a common and widespread Hawaiian *Asplenium*, and its true nature had been overlooked by pteridologists. The important point is that this plant had the predicted features of the hypothetical ancestor of *Diellia*, and although it evidently became extinct (for it was never found again), it can, for all practical purposes, be counted as truly a "living ancestor."[46]

The interesting thing here is the notion of "prediction" of past events. Normally, one thinks of prediction as relating to events or phenomena that are ongoing, repeatable, or yet to occur—that is, having significant future manifestations. This normal kind of prediction does have a place in cladistics, and it will be useful to clarify how it differs from yet still connects with the prediction of the past. The taxonomist P. E. Griffiths argues that cladistic explanations of zoological traits are better than the alternative—functional-adaptive ones— because they "classify in a 'maximally predictive' way. They group organisms in a way that best predicts previously unnoticed patterns of resemblance." Crocodiles, for example, bear a close genetic relationship to birds and so exhibit a remarkable array of birdlike characteristics: nest building, brood care, and complex vocalization such as the "peeping" of hatchlings. Because none of these traits is found among other reptiles, zoologists predict that further study will uncover additional similarities shared exclusively by existing crocodiles and birds.[47]

Yet Griffiths's prediction of future discoveries, in this case involving the behavioral traits of living species, does not function in the future dimension alone. It is based upon a contrasting and indeed prior kind of prediction, one concerned with ancestral kinship. Griffiths's expectation of future discoveries is therefore oriented toward the past as much as toward the future. Paradoxical though it sounds, this cladistic mode of prediction concerns things that have already taken place: it tries to recover data about the "original" conditions that must, on the genealogical principle, be called on to account for certain present-day phenomena. Warren Wagner appropriately calls this the "prediction of ancestors."

> Prediction of ancestors has an appeal to scientists in general, because it demonstrates that systematics has a logical structure sufficiently coherent that the researcher can estimate the nature of previously unknown organisms. . . . To be able to postulate the course of evolution in a group of organisms on the basis of indirect evidence is one of the goals of phylogenetic research, since in many groups direct evidence is difficult or impossible to obtain. Conclusions are subject to various degrees of confirmation or falsification by the discovery of new evidence.

Here, no doubt, Wagner has in mind his own discovery that the fern sample from Maui was in fact the "living" reality correlate to his hypothesized ancestor. Surely this had been a defining experience in his career, giving him confidence that similar empirical confirmations could be expected in the future. He adds in this context: "Although an ancestor itself may not be discovered, a form divergent from it but closer to it than any other may be found." The classic case in linguistics was the identification of Sanskrit as the eldest known sister within the Indo-European family.[48]

With this recent cladistic logic in mind, may we predict that such hypothesizing occurred as well in the late nineteenth century? The pattern not only is confirmed but, more importantly, is confirmed in both the biological and philological fields of that period. In 1884, for example, the physiologist William B. Carpenter described his investigation of a certain class of mollusks. He had begun years prior to that by constructing an evolutionary pedigree for these organisms, and now he was able to report the triumphant outcome: "This hypothetical pedigree has found its complete confirmation in a deep-sea Orbitolite of extraordinary delicacy and beauty, which was brought up in the *Porcupine* Expedition of 1869." The purpose behind Carpenter's research gives his finding additional significance: he was trying to discover directional patterns in the evolution of small sea creatures and, in this way, to revitalize the thesis that selection occurred according to design.[49] Such unorthodoxy among transmutationists appeared in several different guises. Yet these philosophical deviations serve only to highlight their opposite: the consensus around common descent that bound these thinkers all back to Darwin.

Other examples of retrospective prediction appear in T. H. Huxley's writings. As early as the 1860s Huxley was pointing out that the search for missing links in the animal kingdom's pedigree would be greatly advanced by any fossil discoveries that narrowed the gap between reptiles and birds. Were there, he asked, any fossil reptiles more birdlike than the reptiles that now exist? Were any fossil birds more reptilelike than the birds of today? These questions amounted to a prediction: hypothesizing that birds had evolved from reptiles, the researcher would forecast that at least some fossils would display characteristics intermediate between these two forms. Confirmation came first with the appearance of the *Archaeopteryx*, having feathers like a bird, a reptilian body, and a mouth with teeth rather than a beak. Later, the Yale paleontologist O.C. Marsh found fossils of similar toothed birds in America. Such findings, Huxley said, were just what the doctrine of descent "would lead us to expect." More complete evolutionary sequences could be reconstructed in a few cases, most notably the supposed ancestry of the horse: paleontologists constructed protoforms of hoofs and other features and found fossilized skeletons of what appeared to be early and highly generalized ancestors. Apparently, paleontology was fulfilling the predictions implied by Darwin's hypothesis.[50]

F. W. Farrar described the linguistic counterpart to this kind of inquiry, an approach made possible by Grimm's law of phonetic change. This principle, Farrar said, "enables us to predict with certainty the form that a [modern] German word, for instance, will assume in the Sanskrit, Greek, Latin, German, Celtic, or Slavonic languages. Thus the English word *queen*, the old Slavonic *zena*, and the Greek *gune* are shown to be identical with the Sanskrit *jani*, a mother."[51] Although his list of languages included only ones that were attested, Farrar was right to call this a kind of prediction: comparison of relatively recent

attested languages, aided by a law tabulating the pattern of phonetic correspondences usually found among them, yielded reconstructed (predicted) words
that could then be checked against the actual words in the oldest known
language.

The most complete programmatic statement of this thesis invoked the
practice of comparative reconstruction in both natural history and philology. In
his *Doctrine of Descent and Darwinism* (1875), the University of Strasbourg zoologist Oscar Schmidt (1823–86) declared that, "down to the minutest details,
linguistic research stumbles on accordance and analogies with the doctrine of
the derivation of organisms."[52] Like Haeckel's books, Schmidt's was replete
with evolutionary pedigree diagrams. Schmidt weighed, moreover, the value of
retrospective hypothesizing in the promotion of biological research. "The
champions of the doctrine of Descent are blamed for often speaking of mere
probabilities, forgetting that even in cases in which the probability ultimately
proves false, the refuted hypothesis has led to progress." The creation and
testing of hypotheses was thus fundamental to scientific progress. And this
openness to "mere probabilities" was not found in biological science only. "Of
this [progress] the science of language has recently borne testimony. It is well
known that linguistic comparison within the family of Indo-Germanic tongues
suggested the reconstruction of the primitive language which formed their
common basis."[53]

Schmidt then added what at first appears to be damaging to the analogy, for
he described a new and convincing critique of the linguistic *Stammbaum* theory.
He referred to the work of Johannes Schmidt (1843–1901)—no relation—a former student of August Schleicher, who had shown that the components of his
teacher's reconstructed Indo-European prototongue actually had originated at
widely different periods, making the primitive language as a whole "a scientific
fiction." By the early 1870s, the leading edge of European philology would
agree that Schleicher's entire *Stammbaum* was flawed and that the neat cleav-
ages it implied were misleading.

Yet as Oscar Schmidt pointed out, Schleicher's mistakes produced an ironic
result. "Nevertheless, inquiry was essentially facilitated by this fiction, and with
it was intimately connected the formation of a pedigree of the Indo-Germanic
linguistic family, as a hypothesis supported by many indications." In other
words, if Schleicher's particular pedigree was untenable, the family tree concept had still proved valuable in stimulating research. "The value of the hypothesis is undiminished. It was the road to truth."[54]

Oscar Schmidt went on to suggest that, despite philology's mistakes,
the language-species analogy still held good. Although he did not mention

Schleicher by name, Schmidt did name Schleicher's intellectual counterpart when he made the comparison: "In our science Haeckel had made the most extensive use of the right of devising hypothetical pedigrees as landmarks for research. It matters nothing that he has repeatedly been obliged to correct himself, or that others have frequently corrected him; the influence of these pedigrees on the progress of the zoology of Descent is manifest to all who survey the field of science."[55] Hence the work of genealogical reconstruction, even if producing mere fictions, was still useful as a hypothesis advancing research.

In sum, an interlocking cluster of themes pervaded both Darwinian natural history and comparative linguistic studies. Moving from the general to the specific, these included: science as a matter of hypothesizing based on probabilities; this hypothesizing having a backward-looking dimension; this backward-look cast in the form of retrospective prediction and future confirmation; and, most specific of all, the reconstruction of lost ancestors. Variations on these themes in the human and cultural disciplines now follow.

Ramifications in the Human Sciences

The intertwined concepts of branching descent, genealogical taxonomy, comparative methodology, and the hypothetical reconstruction of pedigrees appeared in a kaleidoscopic array of language- and text-based disciplines.[56] Here may be included some fields that were not strictly genealogical yet still purported to recover the earliest stages of a developmental process.

One variation appeared in biblical scholarship. Since the late seventeenth century, a stream of intrepid writers had argued that the Pentateuch, the first five books of the Hebrew Bible, was a product of composite authorship. Julius Wellhausen (1844–1918) codified this "documentary hypothesis" in his *Geschichte Israels*, volume 1 (1878)—immediately summarized for English readers in the *Encyclopaedia Britannica*.[57] Wellhausen identified four separate sources, no longer extant, woven together to form the unified biblical narrative. This was of course the exact reverse of the branching pattern, yet it did rely on extrapolation from present sources to reconstruct a vanished past. Conversely, a truly branching genealogy emerged in New Testament studies, in the attempt to discover the developmental relationship between the three synoptic gospels. Many scholars posited a lost text, identified as Q (from the German *Quelle*, "source"), to account for material common to the books of Matthew and Luke yet absent from Mark—hence a hypothesized ancestor.[58] Although controversial, these tendencies in biblical scholarship helped make familiar the notion of controlled reconstruction of extinct predecessors.

Similar patterns emerged in other fields. Manuscript criticism, for instance, required the removal of corruptions found in medieval copies of ancient texts; for this purpose, scholars used a comparative method to reconstruct the lost textual "prototype."[59] More celebrated was the nineteenth century's astounding successes in the decipherment of old writing systems, notably the Egyptian hieroglyphs and Mesopotamia's cuneiform. The latter not only allowed scholars to read the Semitic languages of the Babylonians and Assyrians but also made them aware of an even earlier, non-Semitic, tongue that had also been written in that script. The language was identified as Sumerian, although the existence of a long-forgotten Sumerian civilization was in doubt until the discovery of additional inscriptions brought confirmation. Somewhat later, the decipherment of the Hittite script confounded scholars (including A. H. Sayce), until repeated tacking between conjectural readings and new corrective evidence brought progress.[60]

A comparative methodology also infused a number of approaches to the historical study of human cultures in the latter half of the nineteenth century. A note of clarification is needed here, however, because histories of Victorian cultural anthropology contend that the genealogical pattern went into eclipse during the 1860s: they tell how the tree of ethnological descent gave way to the ascending staircase of sociocultural evolution, the latter suggesting unilinear progression rather than branching radiation. This change was due to the discovery of the vast antiquity of man, exposing philology's shallow historical reach and showing that it would never be able to discover the ultimate pedigree of nations or their institutions.[61]

Yet the eclipse of this universal genealogizing did not preclude an efflorescence of interest in Aryan genealogy. The Indo-European ethnological idea continued as a staple component of comparative philology texts.[62] This same period also saw the advent of "comparative jurisprudence" in Sir Henry Maine's *Ancient Law* (1861), a work patterned on the comparative-philological principle. The same essential method appeared in Fustel de Coulanges's *La Cité Antique* (1864), which enjoyed immense popularity in its 1873 English translation; moreover, T. Childe Barker recapitulated Fustel's argument in his *Aryan Civilization* (1871).[63] Finally, comparative philology supplied the basis for the study of kinship systems within various societies, an application seen in the work of the American ethnologist Louis Henry Morgan.[64]

Perhaps the best known of these genealogically oriented fields was comparative mythology, formally launched in Max Müller's 1856 lecture by that title. Müller elaborated this idea in addresses he later gave at Westminster Abbey and for the University of Glasgow's Gifford Lectures, and his disciple Sir George Wilson Cox seconded the theme.[65] The germ of this approach to mythology had been anticipated by William Jones, who set forth the field's ethnological basis. "When features of resemblance, too strong to have been accidental, are observable in different systems of polytheism, without fancy or prejudice to colour them and improve the likeness, we can scarce help believing, that some connection has immemorially subsisted between the several nations, who have adopted them."[66] As Jones suggested, comparative mythology involved not just the diffusion of discrete beliefs but implied the splitting of a formerly unified people into separate "nations." An 1863 article in the *Reader* summed up this approach to folklore studies: "Philologists of the Grimm school have traced back to a common source in the Sanskrit language and literature the traditions which both Europeans and West Asian peoples have believed, and the customs they have practiced and clung to."[67]

Another application of comparative methodology appeared in "linguistic paleontology," a field christened by the Swiss linguist Adolph Pictet (1799–1875) in his *Les origines indo-européennes ou les aryas primitifs* (1859, 1863)—the first volume appearing in the same year as *The Origin of Species*. Inspired by the natural historian's pursuit of fossils, Pictet sought to recover a portion of the Indo-European "proto-lexicon." The early-twentieth-century anthropologist V. Gordon Childe explained how linguistic paleontology could reconstruct the life mode of the Aryan mother tribe prior to its dispersion.

> The words and names which recur in a plurality of the separate Indo-European languages, duly transformed in accordance with . . . phonetic laws . . . , constitute in their totality the surviving vocabulary of the original Aryans. The objects and concepts denoted by those words are therefore the objects and concepts familiar to the ancestors of the Indo-European peoples. The sum of such corresponding terms would then depict the culture of the primitive people.[68]

Linguistic paleontology was quickly absorbed into the mainstream of nineteenth-century comparative philology, and it continued into the twentieth century in more detailed form through Otto Schrader's *Prehistoric Antiquities of the Aryan Peoples* (1890) and Childe's *Aryans: A Study of Indo-European Origins* (1926).[69] These works reviewed the various theories concerning the details of Indo-European linguistic kinship and migration patterns; to illustrate these, they reproduced the family tree diagrams constructed by Schleicher, Friedrich Müller, August Fick, and others (see fig. 5.16). While the project of linguistic paleontology always provoked controversy, it has persisted and it underwent considerable technical elaboration.[70] Its multiple versions have continued to reinforce the notion of genealogical reconstruction in the human sciences.

Comparative philology was at first based on the assumption that migrations constituted the reality underlying the family tree of Indo-European myth and

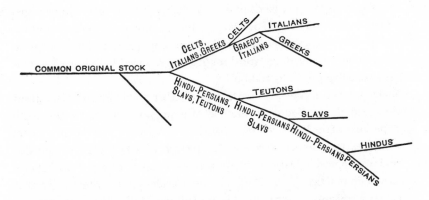

Fig. 5.16. F. Müller, pedigree of Indo-European nations, reprinted in Schrader, *Prehistoric Antiquities of the Aryan Peoples* (1890).

language. On this view, the emergence of new, mutually unintelligible languages required the decisive splitting of a previously unified speech community. Yet by the 1870s, as was noted earlier, both the migration model and the related *Stammbaum* concept had been discredited. Detractors pointed out the lack of scholarly consensus as to the order in which groups departed from the main ethic stock and formed new tribes, resulting in conflicting versions of the Indo-European tree of descent. More fundamentally, critics faulted the *Stammbaum* for suggesting too clear-cut a picture of tribal and linguistic division in the first place. Philologists subsequently produced evidence of dialectical differentiation even in the absence of the sharp splitting of populations through migration, of slight linguistic differences spread over large populations which were still more or less united in speech.[71]

Yet the new understandings of actual on-the-ground linguistic change in a sense had little effect, and the *Stammbaum* retained its popularity. First of all, at the very time that the genealogical model was being faulted, the methodology to which it was wedded triumphed in the work of the Neogrammarians.[72] Moreover, the *Stammbaum* has remained a convenient shorthand for summarizing at a glance historical-linguistic relationships. Its diagrammatic manifestations continue to appear at present, even in works directly critical of the concept. And even in these, linguistic pedigrees have not served as mere foils for attack but have continued in their role as heuristic devices, conveying essential information about linguistic kinship.[73]

To the extent that the idea of ethnic migration itself persisted, as it did in some versions of Aryan studies, it strongly paralleled Darwin's notion of geographic speciation. In the latter schema, physically isolated populations, such as the mockingbirds on the islands of the Galapágos chain, evolve into new species in the course of time. For Darwin, this phenomenon pointed toward divergent evolution; it helped explain the branching tree of life, just as—as he had seen in his own reading—tribal migration grounded Indo-European ethnology's genealogical configuration.[74]

Finally, let us consider the question of whether philology and Darwinism shared a hypothetico-deductive method of inquiry. Confusion on this topic reigns because of the longstanding debate between historians and philosophers of science as to whether Karl Popper's picture of a hypothetico-deductive method was applied in Darwin's work on transmutation, or whether it reflects real scientific practice at all. These questions aside, some writers argue that genealogical reconstructions, such as are still used in biological systematics today, constitute testable hypotheses and indeed offer the only approach yielding propositions open to falsification. Moreover, some also argue that comparative linguistics makes analogous use of cladistic method, including a Popperian approach to inquiry.[75]

Without taking sides on these thorny questions, we can at least note that twentieth-century language study has produced two dramatic confirmations of a "predicted ancestor." Or more precisely, events have confirmed retrospective hypotheses originally deduced from linguistic laws. The first case involved Ferdinand de Saussure's *Mémoire sur le système primitif des voyelles dans les langues indo-euopéennes* (1879), which reconstructed the proto-Indo-European vowel system. Saussure's hypothesis was confirmed after his death by the decipherment of the Hittite script used in Anatolia (modern Turkey) during the second millennium B.C.E. The first step came when researchers concluded that the Anatolian tongues had been part of the Indo-European family and that they had separated earliest from the main Indo-European stock. Then, in the 1930s, they identified Anatolian "laryngeals," sounds absent from all subsequent Indo-European languages yet coordinate to the pattern Saussure had forecast half a century earlier.[76]

A similar case involved ancient Greek. Nineteenth-century Hellenists conjectured the early features of the language. Then British archaeologists discovered in Crete and southern Greece a hitherto unknown script, the celebrated Linear B. When finally deciphered in the 1950s, Linear B revealed the Mycenean Greek of the thirteenth century B.C.E., a form of the language several hundred years older than any previously attested. This find confirmed the predictions of earlier scholars: certain grammatical features, once postulated only in reconstructed forms, appeared where expected.[77]

Confirmations of this kind are necessarily rare, for they depend on a largely fortuitous combination of events: a hypothetical reconstruction followed by new discoveries about the earliest stages of the relevant language group. And for the purposes of historical argument—capturing the logic shared in that earlier period by philology and Darwinism—these more recent finds are not really significant: they were too technical and appeared too late to have produced an impression on readers which significantly reinforced analogous Darwinian conjectures. Still, the less dramatic linguistic confirmations of the late nineteenth century would have exerted their own influence, as the linguist W. S. Allan explains. "If the [newly] documented history of a language supports a reconstructed form, this implies not only that the result in this case is correct, but that the methodology in general is correct, and may be applied with some confidence to languages which have little, if any, documented history."[78] This pattern of discovery, in whichever form it appeared, would have strengthened the sense of the scientific rigor of comparative-historical hypothesizing.

Epilogue

Conjectural Genealogical Reconstruction, Antiquarian Aesthetics, and the Plausibility of Common Descent

IN EXAMINING THE eighty years after the *Origin* appeared, historians have tended to take the descent idea for granted as the generally accepted aspect of Darwinism and to focus instead on the debates over natural selection. Here I ask why common descent itself became so widely accepted, particularly among a lay readership. This outcome was of course mainly due to the scientific arguments made by Darwin and his followers and to the confirmations supplied by subsequent research. Yet the very plausibility of this substantive research, I suggest, was buttressed by extrinsic factors, especially by the way in which evolutionism fitted with a kind of reasoning that permeated large sectors of the scholarly world. Of course, a similar argument is sometimes made in terms of Darwinism echoing the nineteenth-century *Zeitgeist* of naturalistic and process-oriented philosophies. Yet I have suggested that Darwin's theory derived additional stature from something more specific and concrete, its fairly precise methodological convergence with particular intellectual disciplines.

Approaching the question from the opposite angle, one may ask what was the probable impact of that disciplinary convergence on readers. How were they affected who, for instance, browsed through books on evolution as well as encyclopedias or dictionaries and saw the parallel kinds of genealogies there represented? Did this tacit concurrence between fields serve the same function as the one hoped for by writers who employed the language-species analogy itself? That is, did it enhance the "naturalness" of Darwinism's picture of common descent? I have tried to show that the answer is yes, that the sheer abundance of genealogical themes in the linguistic fields must have corroborated and reinforced the Darwinian idea of descent, thus increasing its intellectual weight and plausibility. This impression would have suggested itself, at least on a half-conscious level, even when no one was pointing it out.

What readers did need to gain was an overall impression of confidence in the techniques of genealogical reconstruction. And this was in fact the single most outstanding achievement of comparative-historical philology since the middle of the nineteenth century, its tremendous success in making intelligible the obscure linguistic past. This was certainly the impression that philology made on contemporary observers among both practicing scholars and the lay public, most authoritatively articulated for English readers in Max Müller's pronouncements. And if the comparative-linguistic fields were in many ways different from paleobiology, even well-informed lay people were as dependent on expert testimony in one of these domains as in the other. That is, they were no less dependent on the philologist's esoteric knowledge with respect to hypothetical reconstructions than they were on the naturalist's.

Moreover, readers need not have thought their way through to a rational acceptance of this disciplinary parallelism. They needed only the sense that two very different kinds of research were coming up with the same general conclusions, such that the scholarly whole appeared greater than the sum of its parts. It is enough, then, to conclude that the multiple permutations of genealogical reconstruction in the human sciences strongly resonated with Darwinism, helping habituate informed audiences to thinking in such terms.

This kind of argument, of course, admits to only the most impressionistic demonstration. And this is perhaps especially true when the question at issue is a fairly technical one on which most educated people nevertheless end up forming some opinion. The plausibility of my own argument can be gauged only by the impression made on present readers by the material set forth in chapter 5. Through the weight of these examples, I have tried to re-create the intellectual consensus our encyclopedia browser or casual reader of *Nature*, the *Athenaeum*, or the *North American Review* would have stumbled upon.

The larger question here concerns how people adopt or change their opinions, often in the absence of thoroughly reasoned reflection. I suspect that at least part of this process is a matter of aesthetics, an overlooked realm in the history, although perhaps not the philosophy, of science. One finds attractive not only the elegance that comes with simplicity of explanation but also the harmony produced by similar ideas in widely different departments of knowledge. The urge to systematize knowledge, to find organizing principles, has of course always been with us. Yet it is one thing for this to be imposed by the mind of an Aquinas and another thing for such intellectual congruity to well up, apparently, from within the structure of knowledge itself. The latter case is the more satisfying. This is not only a matter of being reassured that the experts, the circle of authoritative inquirers, have reached a degree of consensus. The aesthetic aspect is more a matter of the world itself seeming harmonious and intelligible. One senses that the first kind of intellectual unity, while desireable, is a sociological construct, while the second seems to adhere in the very nature of things.

This second condition is the real attraction, after all, of analogy and metaphor, how they embody not only a logic of argumentation but an aesthetic of similitude that borrows from a foreign quarter for its inspiration. An analogy works because of its combined similarity and difference, that combination allowing the beholder to see the same thing in a new way. A separate, contrasting, and in a sense foreign tradition of inquiry, such as philology represented, therefore provided an effective complement to natural science, validating Darwinian concepts as constituents of a greater ideal whole.

Comparative philology also had the inherent benefit of combining a scientific method and a romanticist, antiquarian spirit, an association that could have emerged so prominently perhaps only in the nineteenth century. The antiquarian ethos, a vast topic in itself, united much of that era's scholarship, transcending boundaries between the sciences and the humanities. It showed itself in a varied array of concrete pursuits: in collecting and arranging, in the classifying of things in ordered *taxa*, in fossil hunting and in stocking geological cabinets; in the interest in old manuscripts and etymologies, in numismatics and inscriptions, in origin myths and buried cities; in the periodization of style in architecture, sculpture, and painting; in recovering lost civilizations and deciphering forgotten writing systems; in metaphors of treelike growth, in tracing one's own family lineage. From the enthusiasm for natural history to the rage for discovering (and often inventing) national origins, in the irresistible analogy between archaeology and paleontology, in all of these fields, a historical consciousness pervaded. From this unifying aesthetic perspective, science and history—the reconstruction of the past—were not at antipodes but at one. This of course did not exhaust the definition of science, for there were other natural-scientific domains besides natural history. But it did mean that science did not exclude retrospection and even reserved for it a valued place. By embodying in themselves so many of these themes, philology and its allied disciplines helped construct the scaffolding of plausibility surrounding the house of Darwin.

NOTES

Prologue. Science as Indirect Discourse

1. A useful guide to this literature appears in Nancy Leys Stepan, "Race and Gender: The Role of Analogy in Science," *Isis* 17 (1986): 261–77. For exploration of the theme in Darwin studies, see Gillian Beer, *Darwin's Plots: Evolutionary Narrative in Darwin, George Eliot and Nineteenth-Century Fiction* (Boston: Routledge & Kegan Paul, 1983), 79–103; L. T. Evans, "Darwin's Use of the Analogy between Artificial and Natural Selection," *Journal of the History of Biology* 17 (1984): 113–40; Robert M. Young, *Darwin's Metaphor: Nature's Place in Victorian Culture* (Cambridge: Cambridge University Press, 1985). The examples could be multiplied.

2. The issue of what fields of knowledge an age assumes naturally go together could use more original exploration by historians of science. Michel Foucault, *The Order of Things: An Archeology of the Human Sciences* (New York: Pantheon, 1970), gives a bold and by now nearly classic treatment, yet his conclusions ought to be reexamined.

3. The existing philosophical literature on aesthetics in science deals mainly with the preference for simplicity and economy in conducting scientific inquiry, not the aesthetics of congruency among larger structures of knowledge as a factor in the presentation and public acceptance of scientific conclusions. See, for instance, the introduction to Nicholas Rescher, ed., *Aesthetic Factors in Natural Science* (Lanham, Md.: University Press of America, 1990); James W. McAllister, *Beauty and Revolution in Science* (Ithaca, N.Y.: Cornell University Press, 1996).

4. Aspects of this topic are treated in Alvar Ellegård, *Darwin and the General Reader: The Reception of Darwin's Theory of Evolution in the British Periodical Press, 1859–1872* (1958; reprint, Chicago: University of Chicago Press, 1990), 291–92; Gillian Beer, "Darwin and the Growth of Language Theory," in *Nature Transfigured: Science and Literature, 1700–1900*, ed. John Christie and Sally Shuttleworth (New York: Manchester University Press, 1989), 152–70; Robert J. Richards, *Darwin and the Emergence of Evolutionary Theories of Mind and Behavior* (Chicago: University of Chicago Press, 1987), 200–206. Relevant attention to Lyell appears in M. S. J. Rudwick, "Historical Analogies in the Geological Works of Charles Lyell," *Janus* 64 (1977): 96–97; Rudwick, "Transposed Concepts from the Human Sciences in the Early Work of Charles Lyell," in *Images of the Earth*, ed. L. J. Jordanova and Roy S. Porter (London: British Society for the History of Science, 1979): 72–73; Thomas R. Trautmann, *Lewis Henry Morgan and the Invention of Kinship* (Berkeley: University of California Press, 1987), 215 n. 19; Trautmann, *Aryans and British India* (Berkeley: University of California Press, 1997), 56–57; Liba Taub, "Evolutionary Ideas and 'Empirical' Methods: The Analogy between Language and Species in Works by Lyell and Schleicher," *British Journal for the History of Science* 26 (1993): 171–93.

5. Stuart Piggott, *Ruins in a Land Scape: Essays in Antiquarianism* (Edinburgh: Edinburgh University Press, 1976), 6–8, 20–21.

6. Historians have at times blended these two ideas rather than distinguished them: see Robert Nisbet, "Genealogy, Growth, and Other Metaphors," *New Literary History* 1 (Spring 1970): 351–63.

7. S. F. Cannon, preface, to Peter Vorzimmer, *Charles Darwin: The Years of Controversy* (Philadelphia: Temple University Press, 1970), xiv.

8. F. Max Müller, *Lectures on the Science of Language,* 2 vols. (1862, 1863; reprint, New York, 1865), 1:354. The idea of linguistics functioning as a kind of natural theology is discussed in J. W. Burrow, "The Uses of Philology in Victorian England," in *Ideas and Institutions of Victorian Britain,* ed. Robert Robson (London: G. Bell & Sons, 1962), 180–204; Hans Aarsleff, *The Study of Language in England, 1780–1860* (Princeton: Princeton University Press, 1967), 182–85, 209, 212, 225–26; Susan Jeffords, "The Knowledge of Words: The Evolution of Language and Biology in Nineteenth-Century Thought," *Centennial Review* 31 (1987): 66–83.

9. Peter Bowler makes this point in his "Darwin on Man in *The Origin of Species:* A Reply to Carl Bajema," *Journal of the History of Biology* 22 (1989): 499.

10. Martin Rudwick, introduction to Charles Lyell, *Principles of Geology,* 1st ed., ed. Rudwick, 3 vols. (1830–33; facsimile, Chicago: University of Chicago Press, 1990), 1:xviii–xix; unless otherwise stated, all further references are to this edition. Gillian Beer makes a similar point concerning Darwin, who likewise dealt with a change process that could not be observed directly: "It is this historical, or proto-historical, element in his work which means that he must give primacy to imagination, to the perception of analogies." Beer, *Darwin's Plots,* 98. This breaking of the "imagination" barrier goes beyond the function of Darwin's metaphors discussed by the philosopher of science Edward Manier. In Manier's terms, Darwin's linguistic figures performed fairly straightforward "critical [i.e., polemical]-persuasive" and "heuristic" functions. The latter entailed conceptually tying together discrete themes, in this case gradualism and development with divergence and taxonomy. See Manier, *The Young Darwin and His Cultural Circle* (Dordrecht-Holland: D. Reidel, 1978), 181–86.

11. Ernst Mayr, "Darwin's Five Theories of Evolution," in *The Darwinian Heritage,* ed. David Kohn (Princeton: Princeton University Press, 1985), 755–72; Mayr, *One Long Argument: Charles Darwin and the Genesis of Modern Evolutionary Thought* (Cambridge: Harvard University Press, 1991).

12. See Peter Bowler, *Evolution: The History of an Idea,* rev. ed. (Berkeley: University of California Press, 1989), 129–34, 143, 148, on the contrast between branching and unilinear conceptions of the relationship among species linked through transmutation. This break with the past qualifies Bowler's stress on the continuity in developmentalist theory before and after Darwin's *Origin.* Bowler, *Evolution,* 149–50, 188, 191. It also means that we need not go so far as to equate the late nineteenth century's eclipse of natural selection with an "eclipse of Darwinism" or with a "non-Darwinian revolution." See Bowler, *The Eclipse of Darwinism* (Baltimore: Johns Hopkins University Press, 1983); Bowler, *The Non-Darwinian Revolution* (Baltimore: Johns Hopkins University Press, 1988); and, for a concise summary of his thesis, Bowler, "Darwinism and Modernism: Genetics, Paleontology, and the Challenge to Progressionism, 1880–1930," in *Modernist Impulses in the Human Sciences, 1870–1930,*

ed. Dorothy Ross (Baltimore: Johns Hopkins University Press, 1994), 236–54. Helpful commentary on this subject appears in Robert J. Richards, *The Meaning of Evolution: The Morphological Construction and Idealogical Reconstruction of Darwin's Theory* (Chicago: University of Chicago Press, 1992), 113–14 and elsewhere.

13. The philosopher Herbert Spencer anticipated Darwin in stating this genealogical thesis. Seeking later to establish his priority as an inventor of modern evolutionism, Spencer rightly noted that his 1857 essay, "Progress: Its Law and Cause," did not anticipate the idea of natural selection but, indeed, set forth "the view that the succession of organic forms is not serial but proceeds by perpetual divergence and re-divergence—that there has been a continual 'divergence of many races from one race': each species being a 'root' from which several other species branch out; and the growth of a tree being thus the implied symbol." Herbert Spencer, "Progress: Its Law and Cause," in his *Essays: Scientific, Political, and Speculative,* 3 vols. (New York, 1901), 1:53 n.

14. Howard E. Gruber, "Darwin's 'Tree of Nature' and Other Images of Wide Scope," in *On Aesthetics in Science,* ed. Judith Wechsler (Cambridge: MIT Press, 1978), 121–40. On the tree image, see also S. S. Schewber, "John Herschel and Charles Darwin: A Study in Parallel Lives," *Journal of the History of Biology* 22 (Spring 1989): 47–59.

Chapter 1. Comparative Philology and Its Natural-Historical Imagery

1. See the Essay on Sources below for a primer on the historiography of linguistics.

2. William Jones, "Third Anniversary Discourse: On the Hindus," in *The Collected Works of Sir William Jones,* ed. Garland Cannon, 13 vols. (1803; reprint, New York: New York University Press, 1993), 3:34.

3. Trautmann, *Aryans,* 13 (see Prologue, n. 4).

4. Schlegel, quoted in Robert H. Robins, *A Short History of Western Linguistics* (Bloomington: Indiana University Press, 1967), 172.

5. John Lyon and Phillip R. Sloan, eds., *From Natural History to the History of Nature* (Notre Dame: University of Notre Dame Press, 1981), 1–4. An English translation of Cuvier's views appeared in his *Lectures on Comparative Anatomy* (London, 1802).

6. F. Bopp, quoted in James H. Stam, *Inquiries into the Origin of Language: The Fate of a Question* (New York: Harper & Row, 1976), 224.

7. F. Bopp, *Über das Conjugationssytem der Sanskrit-sprache in Vergleichung mit jenem der griechischen, lateinischen, persischen, und germanischen Sprache* (Frankfurt, 1816); Bopp, *Vergleichende Grammatik des Sanskrit, Zend, Armenischen, Griechischen, Lateinischen, Litauischen, Altslawischen, Gothischen und Deutschen,* vol. 1 (Berlin, 1833). The complete *Comparative Grammar* appeared in six volumes, 1833–52.

8. John William Donaldson, *The New Cratylus, or Contributions Towards a More Accurate Knowledge of the Greek Language,* 3d ed. (London, 1859), preface, xiii. W. B. Winning, *A Manual of Comparative Philology* (London, 1838), 52, says that "philologists are able to conceive an ideal original grammar, from which all the Indo-European idioms have deviated."

9. A. Schleicher, "Die ersten Spaltungen des indogermanischen Urvolkes," *Allgemeine Monatsschrift für Wissenschaft und Literatur* (Kiel) (Sept. 1853): 787; Schleicher, *Die Deutsche Sprache* (Stuttgart, 1859), 82.

10. In *The Naturalist in Britain: A Social History* (London: Allen Lane, 1976), 53–54, David Elliston Allen suggests that the Victorian connection between nature and language emerged from a shared aesthetic ethos, rooted in Romanticism. Allen also notes (13, 59–60) that the naturalists of that era were generally members of the "clerisy" and so were commonly educated in the classical languages.

11. *Dictionary of National Biography*, s.v. "Wedgwood, Hensleigh." Wedgwood was one of the first to write an English-language review of Jacob Grimm's *Deutsche Grammatik*: "Grimm on the Indo-European Languages," *Quarterly Review* 50 (1833): 169–89.

12. Lyell's work was polemically laced not so much because of its actualism, as Lyell himself implied, but more because of the case it made for a steady-state view of the earth's history. See chap. 2, below, for discussion of this issue.

13. The text of Herschel's 1836 letter appears in S. F. Cannon, "The Impact of Uniformitarianism: Two Letters from John Herschel to Charles Lyell, 1836–1837," *Proceedings of the American Philosophical Society* 105 (1961): quotation at 307–8. The Geological Society of London published a portion of this letter in its *Proceedings* in 1837, although this did not contain the linguo-geological reference. Canon, "The Impact of Uniformitarianism," 304; Michael Ruse, "Charles Lyell and the Philosophers of Science," *British Journal of the History of Science* 9 (1976): 121–31.

14. C. Darwin to Caroline Darwin, 27 Feb. 1837, *The Correspondence of Charles Darwin*, ed. Frederick Burkhardt and Sydney Smith, 9 vols. (New York: Cambridge University Press, 1985–95), 2:8–9; C. Darwin to J. M. Rodwell, 5 Nov. [1860], 8:464.

15. William Whewell, *History of the Inductive Sciences*, 3 vols. (London, 1837), 3:482. Whewell also discussed this class of sciences in his *Philosophy of the Inductive Sciences* (1840). Here he explained the composite designation he had invented: "While *Palæontology* describes the beings which have lived in former ages without investigating their causes, and *Ætiology* treats of causes without distinguishing historical from mechanical causation; *Palætiology* is a combination of the two sciences." Whewell, *The Philosophy of the Inductive Sciences, Founded Upon Their History*, 2d. ed., 2 vols. (1847; facsimile, New York: Johnson Reprint Corp., 1967), 1:637–38. Whewell acknowledged that the palaetiological sciences should be divided into those, like geology, which deal with material things, and those "respecting the products which result from man's imaginative and social endowments," such as language (641–42).

16. Whewell, *History*, 484. In its immediate context, this was part of Whewell's affirmation of directionalism in the earth's and life's history and a rejection of Charles Lyell's recurrent, steady-state interpretation of the fossil record.

17. Ibid., 3:482.

18. Ibid., 484. This of course was not yet the language-species analogy that Darwin would introduce, for living species had not yet been added to the mix. Yet Whewell's palaetiological sciences did include paleontology, the field in which geology intersected the life sciences: even if the fossil record were regarded as a succession of discrete events, it was only a short step toward connecting them together and so connecting the language aspect along with it. Whewell moved in this direction in his opposition to Lyell, for he implicitly connected, through his palaetiology concept, comparative linguistic study with geological progressionism, the belief that the fossil record showed a unidirectional trend from simple to more complex creatures. This view would gain ground in the 1850s.

19. Winning, *Comparative Philology*, 4; Donaldson, *New Cratylus*, 12, italics in original.

Chapter 2. From the Early Notebooks to The Origin of Species

1. These two linguistic-origins theories had been popular at least since the seventeenth century. Each would contribute to Darwin's views on the origin of language. See James Burnett-Monboddo, *Of the Origin and Progress of Language*, 6 vols. (Edinburgh, 1774–92). The full title of William Gardiner's book was *The Music of Nature; or, an attempt to prove that what is passionate and pleasing in the art of singing, speaking, and performing upon musical instruments, is derived from the sounds of the animated world* (London, 1832).

2. *Charles Darwin's Notebooks, 1836–1844*, ed. Paul H. Barrett, Peter J. Gautrey, Sandra Herbert, David Kohn, and Sydney Smith (Cambridge: Cambridge University Press, 1987), 581. The beginning of Darwin's parenthetical note is clearly marked by a bracket. "Metaphysical" Notebook N, 64–65.

3. Barry G. Gale, *Evolution without Evidence: Charles Darwin and the Origin of Species* (Albuquerque: University of New Mexico Press, 1982), 108–9, notes the "pedagogical" intent expressed in this notebook entry, although he omits Darwin's allusion to the "radical diversity of tongues."

4. *Darwin's Notebooks*, 599 ("Old and Useless Notes," 5). Darwin received his exposure to Monboddo from Benjamin Smart's *Beginnings of a New School of Metaphysics* (1839).

5. Ibid., italics added. Darwin recalled this theme later, in *The Descent of Man and Selection in Relation to Sex*, ed., with an introduction, by John Tyler Bonner and Robert M. May (1871; reprint, Princeton: Princeton University Press, 1981), 182: "According to a large and increasing school of philologists, every language bears the marks of its slow and gradual evolution." The point here was not to draw a language-species analogy, but to suggest evidence that all of humanity's civilized practices had emerged gradually from primitive roots.

6. *Darwin's Notebooks*, 603.

7. John Horne Tooke, *Diversions of Purley*, rev. ed., 2 vols. (London, 1829), 1:100; also quoted in *Darwin's Notebooks*, 603n.

8. Peter J. Vorzimmer, "The Darwin Reading Notebooks (1838–1860)," *Journal of the History of Biology* 10 (Spring 1977): 125. *Asiatic Researches*, 22 vols. (1786; reprint, New Delhi: Cosmo Publications, 1979): 1:vii.

9. See Mayr, "Darwin's Five Theories," 755–72 (see Prologue, n. 11).

10. By *divergence*, I refer to what may be called the *phenomenon* of divergent change, not the Darwinian "*principle* of divergence," the latter indicating a spreading horizon of adaptations among a population, in order to fully utilize a variety of environmental niches. Howard Gruber suggests that Darwin hit upon the idea of treelike divergence in 1837, then rediscovered this insight, in terms of the "principle," sometime after 1844. Gruber, *Darwin on Man* (New York: E. P. Dutton, 1974), 117–18.

11. In the pre-Darwinian evolutionist scheme, organic diversity was produced by differing rates of parallel development up through the same basic hierarchy. Mayr, *One Long Argument*, 19–21 (see Prologue, n. 11); Bowler, *Evolution*, 144–45, 170 (see Prologue, n. 12); Michael Ruse, *The Darwinian Revolution* (Chicago: University of Chicago Press, 1979), 9–10; Malcolm J. Kottler, "Charles Darwin and Alfred Russell Wallace: Two Decades of Debate over Natural Selection," in *Darwinian Heritage*, ed. Kohn, 381–84 (see Prologue, n. 11). According to a number of students of the Darwinian revolution, the natural-historical argument for the unilinear view of evolution was overturned in the 1840s and even more so

in the 1850s. The most advanced researches in embryology, morphology, and paleontology pointed away from simple progression and toward multiple lines of specialized adaptation to various life conditions: the natural history of a given class of animals, for instance, consisted of radiation outward from earlier existing forms. Although the proponents of this more complex view of natural history were not evolutionists, their work was ripe for an evolutionist interpretation. Darwin had long since formed his own branching model of descent by the 1850's, yet he was able to recast these more recent findings in transmutationist terms when he came to write *The Origin of Species* (1859). See Bowler, *Evolution*, 129–134, 143, 148, for a summary of this interpretation. An alternate view appears in Richards, *Meaning of Evolution* (see Prologue, n. 12).

12. As the historian Eric Hobsbawm put it, "Philology was the first science which regarded evolution as its very core." Hobsbawm, *The Age of Revolution: 1789–1848* (New York: New American Library, 1962), 337, quoted in Henry M. Hoenigswald, "Language Family Trees, Topological and Metrical," in *Biological Metaphor and Cladistic Classification*, ed. Hoenigswald and Linda F. Wiener (Philadelphia: University of Pennsylvania Press, 1987), 257.

13. Louis Agassiz is quoted in Colin Patterson, "The Contribution of Paleontology to Teleostean Phylogeny," in *Major Patterns in Vertebrate Evolution*, ed. Max K. Hecht, Peter C. Goody, and Bessie M. Hecht (New York: Plenum Press, 1977), 582.

14. The title of Arthur O. Lovejoy's essay "Herder: Progressionism without Transformism" well epitomizes this distinction. Lovejoy's essay appears in Bentley Glass, Owsei Tomkin, and William L. Straus, Jr., eds., *Forerunners of Darwin: 1745–1859* (Baltimore: Johns Hopkins University Press, 1959). Ernst Mayr similarly highlights the contrast between a theory of actual "descent" and the "temporalized" *scala naturae*, which was the "progressionists'" only innovation. Mayr, *One Long Argument*, 22.

15. The suggestion is made in Gavin de Beer, *Charles Darwin: Evolution by Natural Selection* (London: Thomas Nelson, 1958), 95; and Adrian Desmond and James Moore, *Darwin* (London: Penguin, 1992), 215–16. De Beer provides no evidence for his claim, whereas Desmond and Moore point to possible sources of inspiration: J. F. W. Herschel's 1836 letter to Charles Lyell, suggesting a philology-geology analogy, plus Darwin's close contact with Hensleigh Wedgwood, both happening at roughly the time when Darwin was writing his "transmutation" notebooks. Frank N. Eagerton hints at this same argument in his description of the influence of Alexander von Humboldt's translated *Personal Narrative of Travels to the Equinoctial Regions of the New Continent, during the Years 1799–1804* (London, 1814–29) on Darwin during his final year as a university student. "Darwin . . . could have noticed that there is an interesting parallel between the evolution and geography of languages and of biological species. Languages, like species, spread when a population either expands or otherwise migrates from one area to another, and in new locations both are apt to diverge from those found in the place of origin. . . . In both cases the newly separated entities have affinities that indicate their common origin." Eagerton, "Humboldt, Darwin, and Population," *Journal of the History of Biology* 3 (1970): 330–31.

16. In chap. 5, below, I offer an alternative, post-1859 version of the idea of cultural resonance between Darwinism and philology.

17. Although Darwin continued throughout his career to use the term *rudiment*, this was actually a misnomer from the evolutionist's perspective. The term implied rudimentary anticipations (what Louis Agassiz called "prophetic types") of functioning structures

in yet-to-be-introduced species, as they indeed were viewed in "transcendentalist" morphology. Yet the Darwinians really saw these structures as vestigial forms, survivals of what had come before.

18. Charles Darwin, *The Foundations of the Origin of Species: Two Essays Written in 1842 and 1844*, vol. 10 of *The Works of Charles Darwin*, ed. Paul H. Barrett and R. B. Freeman, 20 vols. (New York: New York University Press, 1987), 176.

19. Long existing only in manuscript, the big species book was finally published as *Charles Darwin's Natural Selection, Being the Second Part of His Big Species Book Written from 1856 to 1858*, ed. Robert C. Stauffer (Cambridge: Cambridge University Press, 1975).

20. Ibid., 262. Darwin broached this potential objection to his theory in an eclectic chapter on natural selection itself. As in the *Origin*, this chapter in the big book ranged over the subjects of natural selection, divergence, and extinction, all summarized through a tree diagram. *Darwin's Natural Selection*, 235 ff.

21. Spencer, "Progress," 1:23, 24 (see Prologue, n. 13). Spencer gave other illustrations (e.g., 34) of this idea of heterogeneity masking its homogeneous beginnings—a crucial theme for Darwin as well.

22. H. Wedgwood to C. Darwin [between 31 Mar. and 29 Sept. 1857], *Correspondence of Charles Darwin*, 6:458–59 (see chap. 1, n. 14). (In that same year, Wedgwood published his *Dictionary of English Etymology*.) At the bottom of the last page of Wedgwood's letter, Darwin penciled three series of etymologically related cognate words, apparently in an effort to reckon their visual effect: "Day and Jour: Dies / Diurnes / Giorno / Journal / Jour"; "Episcopus / Obispo / Bishop"; "vescovo / evesque / eveque" (6:458). Concerning the dating of Wedgwood's letter, my view suggests that Wedgwood wrote to Darwin soon after Darwin had had a chance to read Herbert Spencer's article and to make his inquiry about related words based on that reading.

23. *Darwin's Natural Selection*, 384.

24. According to David M. Stamos, Darwin's language analogy (referring to the passage in the *Origin*, chap. 13, however) suggested that, even if species have no real identity through time ("vertical species"), this does not affect the apparent reality of species as distinct entities at present ("horizontal species"): English is no less real whether or not we can say if it changed in the past or will change in the future. As Stamos further suggests, the notion of evolutionism by its very nature blurs the distinction between species nominalism and species essentialism by making the latter obtain only at given points in time. Stamos, "Was Darwin Really a Species Nominalist?" *Journal of the History of Biology* 29 (Spring 1996): 139–40.

25. *Darwin's Natural Selection*, 384.

26. "Ancient and extinct forms of life are often intermediate in character, like the words of a dead language with respect to its several offshoots or living tongues." C. Darwin, *Variation of Animals and Plants under Domestication*, vols. 19, 20 of *Works of Charles Darwin*, 19:9. Darwin used this simile in pointing out the relatedness between species then existing and fossil species in each geographic region. The larger context was a hasty overview and defense of his descent theory. Later in that book (20: 199–200), Darwin made a further linguistic reference when he noted how "the science of language" demonstrated that domestic breeding had been practiced since the time when "the Sanskrit, Greek, Latin, Celtic, and Slavonic languages had not as yet diverged from their common parent-tongue."

The main point here was to suggest, through real linguistic evidence, the abundant time during which "selection by man" must have taken place and the significant effects it therefore could have produced. Yet the analogy with common descent was implicit as well.

27. "We can understand, on the genealogical view of classification, how it is that systematists have found rudimentary parts as useful as, or even sometimes more useful than, parts of high physiological importance. Rudimentary organs may be compared with the letters in a word, still retained in the spelling, but become useless in the pronunciation, but which serve as a clue in seeking for its derivation." Charles Darwin, *On the Origin of Species* (1st ed., 1859; facsimile, Cambridge: Harvard University Press, 1964), 455.

28. Ibid., 40. In its more specific context, this language analogy culminated Darwin's exposition of his idea of "unconscious selection" (34–40).

29. Beer, "Darwin and the Growth of Language Theory," 162 (see Prologue, n. 4), points to anti-essentialism as what attracted Darwin to linguistic philosophy, especially in that of Dugald Stewart.

30. Spencer, *Essays,* 1:52.

31. Darwin, *Origin,* 280.

32. Ibid., 301.

33. Ibid., 310–11. The bibliographic aspect of Darwin's image resonated with at least one sympathetic reader. A geologist told Darwin that he found by his own observations of fossils "the distinct and unquestionable marks of a lost record, of which all the details are gone. . . . I compare it to Livy's History. We have some chapters tolerably complete. Of others we have the mere heads, but enough to know that they once existed." Joshua Toulmin Smith to C. Darwin, 6 Jan. [1860], *Correspondence of Charles Darwin,* 8:23.

34. Beer similarly argues that the "crucial attraction" of linguistic theory for Darwin was its reliance on the discovery of laws of change in the absence of exhaustive evidence, including among the latter the "irrecoverable" character of the parent language. Beer, "Darwin and the Growth of Language Theory," 159, 163.

35. Darwin wrote just prior to the first effort actually to reconstruct the Indo-European protolanguage: August Schleicher's *Compendium der vergleichenden Grammatik der indogermanischen Sprache* (1861).

36. The leading Victorian scientific philosophers J. F. W. Herschel and William Whewell urged the method of eliminative induction. See Martin Rudwick, "Darwin and Glen Roy: A 'Great Failure' in Scientific Method?" *Studies in the History and Philosophy of Science* 5 (1974): 97–180; Janet Browne, *Charles Darwin: Voyaging* (New York: Knopf, 1995), 1:438, 440.

37. See relevant passages from Lyell's *Elements of Geology* quoted in *Darwin's Notebooks,* 352–53 n. Lyell, *Elements of Geology* (London, 1838), 272, also used the analogy of the history book simply to describe the chronological sequence inherent in horizontal strata: "a lofty pile of chronicles." In a condensed edition, Lyell shortened this analogy yet added a telling elaboration: he noted the "imperfection" of the geological record, declaring that "even in the volumes which are extant the greater number of the pages are missing," while those remaining had "but few and casual entries." Charles Lyell, *The Student's Elements of Geology* (New York, 1871), 273. Lyell retained this same image in his *Geological Evidences of the Antiquity of Man: with Remarks on Theories of the Origin of Species by Variation* (Philadelphia, 1863), 448–49. (Expanded from what was originally a fourth book of Lyell's *Principles of Geology* [1831–33], *Elements* dealt with "geology proper," or past changes in the earth,

whereas *Principles* treated "the economy of existing nature" as illustrating the kinds of change that must have taken place in the past. For a useful explanation of the relations among Lyell's books, see the preface to Lyell, *Principles of Geology*, 10th ed. [London, 1867], vol. 1).

38. *Darwin's Notebooks*, 352–53 (Notebook D, 60). See also ibid., 398 (Notebook E, 5–6) and 433 (Notebook E, 127) Darwin later expressed similar appreciation for one of Lyell's illustrations: "Capital that metaphor of the clock," referring to Lyell's comparison, in his 1850 presidential address of the Geological Society, of the time needed for geological change to a clock's minute hand and the life span of most organic species to the hour hand. Charles Darwin to Charles Lyell, 8 Mar. 1850, *Correspondence of Charles Darwin*, 4:319–20. The clock analogy appears in Lyell, Presidential Address, *Quarterly Journal of the Geological Society of London* 6 (1850), xlvi.

39. Lyell illustrated this pattern, first of all, by describing the effect of a roving population census on the mortality statistics of several mutually distant counties: any comparison of county totals from the same point in time obviously would be invalid. Lyell's point was that the oft found differences between the kinds of fossil remains in adjacent strata were likewise misleading, "a necessary consequence of the existing laws of sedimentary deposition, accompanied by the gradual birth and death of species." Lyell, *Principles*, 3:30–33 (see Prologue, n. 10). For a detailed analysis of Lyell's argument in this passage, see Rudwick, introduction to Lyell, *Principles*, 1:xxxviii–xxxix.

40. Michael Bartholomew, "Huxley's Defence of Darwin," *Annals of Science* 32 (1975): 530; Lyell, *Principles*, 3:33.

41. *Life, Letters and Journals of Sir Charles Lyell*, ed. Katherine Murray Lyell, 2 vols. (London, 1881), 1:214–15.

42. Lyell, *Principles*, 3:33–34.

43. Another linguistic analogy in Lyell's *Principles*, implying true linguistic development, appeared at the end of vol. 1 (460–62). With no reference to *species*, however, this illustration was meant to show that the obscurity of subterranean processes did not disprove the "uniformity" of the earth's system.

44. It is hard to tell whether Lyell believed that the Latin tongue had descended from the ancient Greek in the same way that "Italian" had from Latin. To posit such a relationship between Latin and Greek would have constituted a gross philological error, one dispelled in Lyell's day by the Indo-European principle, which showed Greek and Latin to be siblings within their linguistic family. Franz Bopp's *Conjugationssystem* of 1816 declared that Greek, Latin, and Sanskrit were "subsequent variations of one original tongue," and in this sense coordinate. Bopp, quoted in Otto Jespersen, *Language: Its Nature, Development and Origin* (1922; reprint, London: Allen & Unwin, 1959), 48. In 1836 the German philologist August Pott elaborated: "An opinion prevailed for a long time, but very erroneously, that Latin is a daughter, or at least a derivative, of the Greek language. . . . And as to the opinion that the Aeolic dialect is more ancient than Latin, the direct contrary is much nearer the truth; for though the Roman idiom is not nearly so copious as its classical neighbour, yet it has preserved its structure and inflexions much nearer to the primeaval form." Translated and quoted in 1838 by Winning, *Comparative Philology*, 43–44 (see chap. 1, n. 8). And as Peter Giles noted, once William Jones made clear the relationship of Sanskrit to the classical tongues, scholars saw that "old notions such as that Latin was derived from a dialect of

Greek must be given up. Men now realised clearly that the relation between Greek and Latin was not that of mother and daughter but of sisters." Giles, *A Short Manual of Comparative Philology for Classical Students* (London, 1895), 8.

45. *Encyclopaedia Britannica*, 15th ed. (1966), s.v. "Herculaneum."

46. Since Lyell used whole languages to represent whole species, the replacement of individual words would suggest the gradual character of the birth and death of a single species, its individual members slowly appearing or dying out. As Martin Rudwick notes, Lyell's archaeological illustration depicts both languages and species as gradually changing in this piecemeal way. Rudwick also suggests, however, that Lyell mistakenly omitted the "new" philology's emphasis on the fluid development not only of whole languages but also of individual words, the latter something quite different from mere word replacement. Yet I believe that this omission was not a mistake but was intentional. See Rudwick, "Historical Analogies," 96–97 (see Prologue, n. 1).

47. C. Lyell to T. H. Huxley, 9 Aug. 1862, quoted in Desmond and Moore, *Darwin*, 512.

48. Lyell's archeaology illustration migrated forward through the editions of his *Principles*. In the first edition (1833), it appeared in chap. 3 of vol. 3; in the fifth edition of 1837, it appeared toward the end of vol. 2; in the tenth edition of 1867, it brought a resounding conclusion to book 1 (chap. 14) of vol. 1.

49. Darwin, *Origin*, 130, 129.

50. Ibid., 420–22.

51. Ibid., 422–23.

52. *Charles Darwin's Marginalia*, ed. Mario A. di Gregorio, with the assistance of N. W. Gill (New York: Garland Publishing, 1990), 1:483 (italics in original). R. G. Latham, *Man and His Migrations* (London, 1851), 74–75.

53. Nancy Stepan, *The Idea of Race in Science: Great Britain, 1800–1960* (Hamden, Conn.: Archon Books, 1982), 33–35.

54. Winning, *Comparative Philology*. Donaldson said similarly, "there is in fact no sure way of tracing the history and migrations of the early inhabitants of the world except by means of their languages." Donaldson, *New Cratylus*, 11 (see chap. 1, n. 8).

55. Latham, *Man and His Migrations*, 27.

56. Spencer's 1857 essay "Progress: Its Law and Cause" suggests that this reliance on linguistic evidence of racial kinship was still the accepted view. "There have sprung many now widely-different tribes, which are proved by philological evidence to have had a common origin," and "philology makes it clear that whole groups of races now easily distinguishable from one another, were originally one race." Spencer, *Essays*, 1:18, 52.

57. Peter Bowler notes that the contingencies of migration of biological populations made real historical reconstruction well-nigh impossible, thereby accounting for the abstract character of the argument in the *Origin*. Bowler, *Charles Darwin: The Man and His Influence* (Cambridge, Mass.: Blackwell, 1990), 79–81.

58. *Darwin's Marginalia*, 1:164. Darwin conceded this point later (although referring to individual traits, not whole species), in a letter to his German translator: "I can in no one instance explain the course of modification." C. Darwin to H. G. Bronn, 5 Oct. [1860], *More Letters of Charles Darwin*, ed. Francis Darwin and A. C. Seward, 2 vols. (New York, 1903), 1:172–73.

59. As an American writer representing the anatomically oriented polygenist school said: "We confess that we do not regard [physical] ethnology as sufficiently advanced to

enable us yet to adopt any certain ethnological classification of languages, which of course can only be based upon the actual filiation of races." J. W. Miles, "On the Philosophy of Language," *Southern Quarterly Review* 20 (Oct. 1851): 407.

60. Darwin, *Origin*, 420–21.

61. An additional point to the analogy was probably that such unequal adaptation was not incompatible with his selection theory. Darwin later said, "In judging the theory of Natural Selection, which implies that a form will remain unaltered unless some alteration be to its benefit, is it so very wonderful that some forms should change much slower and much less, and some few should have changed not at all [?]" C. Darwin to G. Bentham, 22 May [1863], *The Life and Letters of Charles Darwin*, ed. Francis Darwin, 3 vols. (New York, 1887), 3:24–25.

62. One might wonder that, to achieve the greatest impact on the reader's imagination, Darwin did not orient his illustration concerning biological taxonomy toward the Indo-European language family, which scholars confidently knew to be derived from an unattested mother tongue. Yet this analogy would have lost the resonance with racial monogenesis, on which Darwin apparently placed a high value.

63. "D. Stewart lives of Adam Smith, Reid, etc. worth reading, as giving abstract of Smith's views" *Darwin's Notebooks*, 559 (Notebook M, 155).

64. Silvan S. Schweber has analyzed the influence of the Scottish political economists (writers he also calls "conjectural historians") on Darwin's ideas about competition among individuals and the division of labor: see Schweber, "The Origin of the *Origin* Revisited," *Journal of the History of Biology* 10 (Fall 1977): 229–316; and Schweber, "Darwin and the Political Economists: Divergence of Character," *Journal of the History of Biology* 13 (Fall 1980): 195–290. Unmentioned, however, is conjectural history in its more fundamental sense, the one implicit in Darwin's every allusion to organic genealogy.

65. For example: C. Darwin to C. Lyell, 8 Mar. 1850, *Correspondence of Charles Darwin*, 4:319.

Chapter 3. The Darwinian Circle and the Post-*Origin* Debate

1. Friedrich Max Müller, *Comparative Mythology* (London, 1856), 22, 24.

2. These works included: Robert Knox, *The Races of Man* (1850); Charles White, *An Account of the Regular Gradation in Man* (1799); Charles Pickering, *The Races of Man; and their Geographical Distribution* (1854); William Lawrence, *Lectures on Comparative Anatomy, Physiology, Zoology, and the Natural History of Man* (1840); Samuel Stanhope Smith, *An Essay on the Causes of the Variety of Complexion in Man* (1810). This last work Darwin read twice, in 1840 and 1844, and he read several of Prichard's works as well. See Vorzimmer, "Darwin Reading Notebooks," 107–53 (see chap. 2, n. 8).

3. We should not regard monogenism as an inspiration to Darwin, in the sense of informing the argument in the *Origin of Species*, for he had long since formulated his basic theory and needed no new conceptual models of descent. Moreover, monogenetic ethnology did not directly anticipate Darwinism: says George Stocking, "What Prichard studied was the 'evolution of varieties' within a single human species created by God." Stocking, *Victorian Anthropology* (New York: Free Press, 1987), 51. Yet ethnology nevertheless offered a strong analogy to Darwinism.

4. Darwin, quoted in John C. Greene, *Science, Ideology, and World View* (Berkeley: University of California Press, 1981), 98. This would have been in either 1839 or 1847, the two occasions on which Darwin read Prichard's work. Vorzimmer, "Darwin Reading Notebooks," 122, 139.

5. *Darwin's Marginalia*, 1:483 (see chap. 2 n. 52). According to Latham, displacement of one human tribe by another "has the following effect. It obliterates those intermediate and transitional forms which connect varieties, and so brings the more extreme cases of difference in geographical contact, and in ethnological contrast." Latham, *Man and His Migrations*, 97 (see chap. 2, n. 52).

6. *Darwin's Marginalia*, 1:603.

7. Greene, *Science, Ideology, and World View*, 99–100. The discussion of race, projected as chap. 6 of the species book, was not written until later.

8. Spencer, "Progress," 1:52 (see Prologue, n. 13). Darwin no doubt saw Spencer's article, which appeared originally in the *Westminster Review*.

9. Latham, *Man and His Migrations*, 78, 82; Stocking, *Victorian Anthropology*, 68. Robert Chambers likewise argued for monogenesis based on linguistic evidence in his *Vestiges of the Natural History of Creation* (1844; reprint, New York: Leicester University Press, 1969), 278, 294, 305, 310. Some scientific thinkers took to an extreme this near identity between languages and the peoples who spoke them, thus producing an equivocal concept of "linguistic race." See William Benjamin Carpenter, "Varieties of Mankind," in *The Cyclopaedia of Anatomy and Physiology*, ed. Robert Bentley Todd, 4 vols. (London, 1852), 4:1345–47; Carpenter, "Ethnology, or the Science of Races," *Edinburgh Review*, Oct. 1848, 247; Spencer, "Progress," 1:24, 52.

10. F. Max Müller, "Ethnology v. Phonology," in C. C. J. Bunsen, *Outlines of the Philosophy of Universal History* (London, 1854), 349.

11. Ibid., 352–53. Thomas R. Trautmann, "The Revolution in Ethnological Time," *Man* 27 (1992): 379–97. The low estimate of human chronology as buttressing the linguistic argument for monogenesis appears clearly in Arthur James Johnes, *Philological Proofs of the Original Unity and Recent Origin of the Human Race* (London, 1846), xv–xvi.

12. In Asa Gray's copy of Darwin's *Origin*, the illustration appealing to linguistic ethnology (422) has an X penciled beside it. While there are no explanatory comments, this apparently was a sign of disapproval. For example, Gray made an X alongside another passage (443), where Darwin observed that "many naturalists" believed that what they called "the Natural System" revealed "the plan of the Creator; but . . . it seems to me that nothing is thus added to our knowledge." This was an issue on which Gray clearly disagreed with Darwin. Yet why would Gray similarly disapprove of Darwin's use of the ethnological analogy? In this case, the X probably indicates Gray's awareness that, because of the revolution in ethnological time, the genealogy of languages could no longer be considered a mirror image of the "pedigree of mankind" (Darwin's phrase). If so, then Gray saw the same pattern as did Charles Lyell, who, in his writings of the 1860s, often noted the vastly different time scales on which languages and races diversified, languages changing much more slowly. Asa Gray Collection, Archives of the Gray Herbarium, Harvard University.

13. Edward Lurie, *Louis Agassiz: A Life in Science* (Chicago: University of Chicago Press, 1960), 252–302, esp. 256.

14. Ibid., 252–302.

15. L. Agassiz, "Natural Provinces of the Animal World and Their Relation to the Different Types of Man," in *Types of Mankind*, ed. J. C. Nott and George R. Gliddon (Philadelphia, 1854), lxxvi.

16. L. Agassiz, "The Diversity of the Origins of the Human Races," *Christian Examiner* 160 (July 1850): 139–40.

17. L. Agassiz, "Natural Provinces," lxxii; Agassiz, "Prefatory Remarks," in J. C. Nott and George R. Gliddon, *Indigenous Races of the Earth* (Philadelphia, 1857), vx.

18. [B. N. Martin], "The Original Unity of the Human Race," *New Englander* 8 (1850): 579. In America, before the appearance of Darwin's *Origin*, biblically orthodox northern Evangelicals accepted the apparent biblical teaching that all of humanity was formed from "one blood," and they were usually against slavery. After the *Origin* came out, however, antievolutionist religious writers found more to admire in Agassiz's stance: "It is quite refreshing to find the prince of naturalists and zoologists earnestly and eloquently protesting against this whole development or evolution theory." "Herbert Spencer's Philosophy; Atheism, Pantheism, and Materialism," *Biblical Repository and Princeton Review* 37 (1865): 267.

Complicating the situation, many of the total number of adherents to racial monogenesis after 1859 were not Darwinians: hence the Evangelical *Princeton Review* wished a pox on the houses of both Darwin and Agassiz. On the other hand, nearly all Darwinians were racial monogenists out of logical consistency: they too opposed Agassiz, though not on religious grounds or primarily because of an opposition to race-based slavery. Said London's *Natural History Review*, "All adherents to the modern doctrine of the 'origin of species' must . . . on its general principles be convinced of the common origin of the varieties of man." Review of *The Origin of the Human Races*, by A. R. Wallace, *Natural History Review*, n.s., 4 (1864): 329.

19. Miles, "Philosophy of Language," 404 (see chap. 2, n. 59); J. W. Miles, *The Student of Philology: annual oration delivered before the literary societies of South Carolina College* (Charleston, 1853), 27–28. On James Warley Miles, see Walter Conser, *God and the Natural World: Religion and Science in Antebellum America* (Columbia: University of South Carolina Press, 1992), 92–95; and E. Brooks Holifield, *The Gentleman Theologians: American Thought in Southern Culture, 1795–1860* (Durham, N.C.: Duke University Press, 1978), 66–71.

A similar view of linguistic unity apart from genealogy could be found at this same time in the writings of Charles Kraitsir, a Bostonian of Hungarian birth and a philosopher of linguistic pedagogy. "All men, however diverse they may become in conflicting passions and interests, have yet the same reason, and the same organs of speech. All men, however distant in place, are yet plunged in a material universe, which makes impressions of an analogous character upon great masses. Languages therefore have a certain unity. Differing superficially, more or less, they begin to resemble each other, as soon as the observer goes beneath the surface." Kraitsir, *The Significance of the Alphabet* (1846), quoted in Mary Lowell Putnam, "Significance of the Alphabet," *North American Review* 68 (1849): 166.

20. Agassiz, "Prefatory Remarks," xiii–xv. See also Lurie, *Agassiz*, 265.

21. Richard Owen, President's Address, *Report of the British Association for the Advancement of Science* (1858): xcii–xciv. Owen also hinted at a linguistic refutation of Agassiz. After criticizing Agassiz's polygenism, Owen noted recent advances in ethnology which proved that humankind had existed much longer than the traditional accounts would suggest (xciv–xcv).

22. For Lyell's presence, see *Sir Charles Lyell's Scientific Journals on the Species Question,* ed. Leonard G. Wilson (New Haven: Yale University Press, 1970), 218.

23. The first such conversation between Darwin and Lyell took place probably in April 1856. L. G. Wilson, introduction to ibid., xliv.

24. *Lyell's Scientific Journals,* 198, 247, 250, 375.

25. Lyell did note the *Origin's* analogy between vestigial organs and archaic word spellings (ibid., 376).

26. C. Lyell to George Ticknor, 29 Nov. 1860, *Life, Letters and Journals of Sir Charles Lyell,* 2:342 (see chap. 2, n. 41). On Lyell's reading of Max Müller, see also *Lyell's Scientific Journals,* 376.

27. *Lyell's Scientific Journals,* 351, 411.

28. Ibid., 429. For Lyell's remarks on race and ethnology, see 173, 245, 264–65, 273, 414, 428–29.

29. Ibid., 87, 197.

30. Ibid., 352. See also 376.

31. Ibid., 425.

32. Philip F. Rehbock, *The Philosophical Naturalists: Themes in Early Nineteenth-Century British Biology* (Madison: University of Wisconsin Press, 1983), 19–24, 76–78. On Lyell having read Baden Powell, see *Lyell's Scientific Journals,* 221.

33. Richard Owen, *On the Nature of Limbs* (London, 1849), 49, 70. Actually, Owen made no absolute distinction between the idealistic and teleological modes. He acknowledged the functionally adaptive features possessed by individual species yet argued that conformity to type was the more all-embracing message of comparative anatomy: it took the larger, comparative view to perceive the morphological analogies among a number of different species (38).

34. *Lyell's Scientific Journals,* 422.

35. Ibid., 424.

36. Ibid., 410, 352.

37. C. Lyell to G. Ticknor, 9 Jan. 1860, *Life, Letters and Journals of Sir Charles Lyell,* 2:329–31.

38. Asa Gray, "Natural Selection not inconsistent with Natural Theology," in his *Darwiniana: Essays and Reviews Pertaining to Darwinism,* ed. A. Hunter Dupree (1876; reprint, Cambridge: Belknap Press, Harvard University Press, 1963), 81. A. Hunter Dupree, *Asa Gray* (Cambridge: Belknap Press, Harvard University Press, 1959), 285–88. Gray also used a language-species analogy, though not in a directly Darwinian way, in his 1862–63 Lowell Lectures on botany. Here Gray noted that human migrations had caused "the principal alterations in the geography of plants" over time, hence the value of botany for history and ethnology: "the naturalization of certain vegetables, like the adoption of a few foreign words into a language of different origin, seems to indicate the source from which a migrating race has sprung, and the route it has pursued." Third series of Lowell Lectures, Asa Gray Collection, box AT, folder 36, Archives of the Gray Herbarium, Harvard University.

39. Gray, "Natural Selection," 118–19, 124–25, 121–22, 128–29. Said Gray in 1861 concerning the nonteleological school of natural theology: "I confess to a strong dislike of Baden Powell's writings. He seems to have had a coarse, materialistic, non-religious mind; at least, he is not the sort of man I should select to illustrate the delicate relations between religion and science." *Life and Letters of Asa Gray,* ed. Jane Loring Gray, 2 vols. (Boston, 1894), 2:464. See also Dupree, *Gray,* 293. Gray's judgment that Baden Powell's idealistic

natural theology was coarse and materialistic hardly seems justified and perhaps was conditioned by the similar logic he saw in Agassiz.

40. Gray, "Natural Selection," 129–31. Darwinism, said Gray, added no unresolved problems to natural theology (such as the problem of evil) which were not there in the first place.

41. Asa Gray to C. Darwin, 22 Mar. 1863, *Life and Letters of Asa Gray*, 2:502.

42. C. Darwin to Asa Gray, 10 Sept. [1860], *Correspondence of Charles Darwin*, 8:350 (see chap. 1, n. 14); Dupree, *Gray*, 298–99; C. Lyell to G. Ticknor, 9 Jan. 1860, *Life, Letters and Journals of Sir Charles Lyell*, 2:329–31; Gray, "Natural Selection," 104; *Lyell's Scientific Journals*, 410.

43. Dupree, *Gray*, 277–83, 300–302. See also Dupree, introduction to Gray, *Darwiniana*, xiv–xv.

44. C. Darwin to Asa Gray, 10 Sept. [1860], *Correspondence of Charles Darwin*, 8:350. Context concerning the Darwin-Gray correspondence on design appears in Desmond and Moore, *Darwin*, 510 (see chap. 2, n. 15).

45. Lewes is perhaps best remembered as the unwed companion of the novelist George Eliot (Marian Evans).

46. G. H. Lewes, "Studies in Animal Life, Chapter IV," *Cornhill Magazine* 1 (Apr. 1860): 447.

47. Ibid. G. H. Lewes's linguistic analogy reappeared verbatim in its book version: *Studies in Animal Life* (New York, 1860), 102–6.

48. T. H. Huxley, *On Our Knowledge of the Causes of the Phenomena of Organic Nature* (London, 1862), 142–43. Darwin caught Huxley's philological error: he crossed out the word *Sanscrit* in his copy of the latter's lectures, knowing that that language was an elder sister, not the parent, of Greek. *Darwin's Marginalia*, 1:425. Huxley corrected this error when he republished his lectures, replacing "Sanscrit" with "one original [language]." T. H. Huxley, *Darwiniana* (1893; reprint, New York: Greenwood Press, 1968), 458. Michael Ruse, *The Darwinian Revolution* (Chicago: University of Chicago Press, 1979), 142, suggests that Huxley's critique of Owen's concept of homology on embryological grounds may have been inspired by philology.

49. Record of meeting of 8 Jan. 1861, *Proceedings of the American Academy of Arts and Sciences* 5 (1860–62): 98. W. W. Goodwin had earned a Ph.D. in classical languages at the University of Göttingen in 1855 and had succeeded Felton in 1860 in the Harvard chair of Greek literature. *Dictionary of American Biography* (New York: C. Scribner's Sons, 1928–58), s.v. "Goodwin, William W."

50. *Proceedings of the American Academy of Arts and Sciences* 5 (1860–62): 102; also, American Academy of Arts and Letters, Records, no. 3, 125, Minutes of 8 Jan. 1861, Boston Athenaeum. Lurie, *Agassiz*, 299. In "Louis Agassiz's Arguments against Darwinism in His Additions to the French Translation of the Essay on Classification," *Journal of the History of Biology* 30 (1997): 121–34, Paul J. Morris makes too sharp a distinction when he suggests that Agassiz's opposition to the Darwinian view of species was only peripheral to his main concern with the divine creation of the higher *taxa*. Actually the issues of species and higher *taxa* came together in Agassiz's rejection of common descent: he implicitly addressed the two issues at once, by analogy, when he dismissed the genetic classification of the Indo-European languages.

51. Agassiz retained an interest in the linguistic analogy, as seen in the copy of Lyell's

Antiquity (1863) that Lyell himself presented him. The margins of the volume are unmarked except in chap. 23, "Languages and Species Compared": vertical slash marks appear alongside passages describing how existing languages have derived from languages no longer existing, and how there are gaps in the textual record, testifying to the gradual nature of linguistic change. (See below for an analysis of Lyell's chapter.) Special Collections, Museum of Comparative Zoology, Harvard University.

52. C. Darwin to C. Lyell, 2 Feb. [1861], *Correspondence of Charles Darwin*, 9:17. The original letter from Gray to Darwin has been lost; Darwin quotes from it, however, in the letter to Lyell cited here. Darwin thanked Gray for his report on the academy meeting, saying he was much amused by Agassiz's linguistic argument. C. Darwin to Asa Gray, 17 Feb. 1861, ibid., 30.

53. C. Darwin to J. D. Hooker, 4 Feb. 1861, ibid., 21; C. Darwin to C. Lyell, 2 Feb. [1861], ibid., 17.

54. Müller, *Lectures on the Science of Language*, 1:361 (see Prologue, n. 8).

55. Hensleigh Wedgwood, *A Dictionary of English Etymology*, 2 vols. (London, 1859–62), 1:ii–iii, Desmond and Moore, *Darwin*, 607; [Frances Julia Wedgwood], "The Origin of Language: The Imitative Theory and Mr. Max Müller's Theory of Phonetic Types," *Macmillan's Magazine* 7 (Nov. 1862): 54–60 (authorship attributed in *The Wellesley Index to Victorian Periodicals, 1824–1900*, ed. Walter E. Houghton, 4 vols. [Toronto: University of Toronto Press, 1966], 1:566, 1132).

56. [F. J. Wedgwood], "Origin of Language," 59. The *bishop* illustration would become something of a commonplace among nineteenth-century philological writers, useful for demonstrating that the most phonetically dissimilar words could be genetically related. Hensleigh Wedgwood redeployed this illustration to demonstrate how the original imitative character of humanity's first words would have become obscured over time: "If English, French, and Spanish were barbarous unwritten languages no one would dream of any relation between *bishop, évêque,* and *vescovo,* all immediate descendants of the Latin *episcopus.*" H. Wedgwood, *On the Origin of Language* (London, 1866), 133. The illustration was later used in William Dwight Whitney, *The Life and Growth of Language: An outline of linguistic science* (New York, 1875), 46.

57. [F. J. Wedgwood], "Origin of Language," 60. One interesting twist is that, just three years earlier, Hensleigh Wedgwood had taken a very different view of the use of the imagination in science, describing it as the source of undisciplined speculations that hindered the search for *verae causae*. H. Wedgwood, *Dictionary of English Etymology*, 1:ii.

58. Asa Gray to C. Darwin, 4 Oct. 1862, *Life and Letters of Asa Gray*, 2:488–89.

59. C. Darwin to Asa Gray, 6 Nov. 1862, *Life and Letters of Charles Darwin*, 2:182–83, italics in original (see chap. 2, n. 61).

60. Max Müller opposed the Enlightenment view of language as a humanly invented institution made up of arbitrary sounds to which meanings were attached by convention. Instead, observing that every object, when struck, rings with its own characteristic sound, he taught that human speech emerged when external stimuli "resonated" in the consciousness for the first time, creating "phonetic impressions." He thereby suggested that humanity's first words were somehow inherent in nature. Müller, *Lectures on the Science of Language*, 1:528–29. Müller's hypothesis of the origin of language was often criticized: e.g., "The Origin of Language," *Westminster Review* 86 (1866): 88–122. See also Thomas Hewitt Key, *Language: Its Origin and Development* (London, 1874), 3–6. Key concurred especially

with the most adamant critic of Max Müller's theory, the American philologist W. D. Whitney.

61. Müller, *Lectures on the Science of Language*, 1:379, italics in original. Additional useful commentary on this section of Müller's book appears in Darwin, *Correspondence*, vol. 10 (1862), 445, n. 2.

62. C. Darwin to Asa Gray, 6 Nov. 1862, *Life and Letters of Charles Darwin*, 2:182–83, italics in original (see chap. 2, n. 61).

63. Asa Gray to C. Darwin, 24 Nov. 1862, ibid., 491. Years before, Gray had remarked in a similar vein: "Abler pens than ours have shown, that the agencies in operation will not account for the origin of any created thing." A. Gray, "Explanations of the Vestiges," *North American Review* 62 (Apr. 1846): 472.

64. Gray's copy of *The Origin of Species* (in the Asa Gray Collection, Archives of the Gray Herbarium, Harvard University) is scored frequently with vertical pencil lines in the margins. Three of Darwin's four linguistic analogies described in chap. 1, above, are thus plainly marked (40, 310, 455).

65. For another instance of Darwin having forgotten something he had written about his own theory, see Frank H. T. Rhodes, "Darwinian Gradualism and Its Limits," *Journal of the History of Biology* 20 (1987): 155–56.

66. Desmond and Moore, *Darwin*, 512–13. The bent of Darwin's interest in language at this time showed in a letter to Hooker: he was impressed by Huxley's new work, *Man's Place in Nature*, but added, "His geology is obscure; and I rather doubt about man's mind and language." C. Darwin to J. D. Hooker, 13 Jan. [1863], Darwin Collection, DAR 115:179, Cambridge University Library. Here Darwin referred to the remarks on language at the conclusion of Huxley's second chapter, "On the Relations of Man to the Lower Animals": "Our reverence for the nobility of manhood will not be lessened by the knowledge, that Man is, in substance and in structure, one with the brutes; for, he alone possesses the marvelous endowment of intelligible and rational speech, whereby, in the secular period of his existence, he has slowly accumulated and organized the experience which is almost wholly lost with the cessation of every individual life in other animals; so that now he stands raised upon it as on a mountain top." T. H. Huxley, *Man's Place in Nature* (1862; reprint, Ann Arbor: University of Michigan Press, 1959), 132. Huxley here made the acquisition of language the distinction between man and beast, whereas Darwin would argue that language had grown out of abilities that animals already possessed.

67. C. Darwin to J. D. Hooker, 4 Nov. [1862], Darwin Collection, DAR 115: 168, Cambridge University Library. Aware of Frances Julia Wedgwood's 1862 article in *Macmillan's* reviewing Max Müller's theory of the origin of language, Darwin recommended that Gray read it. C. Darwin to A. Gray, 23 Nov. [1862], Asa Gray Collection, Archives of Gray Herbarium, Harvard University.

68. Asa Gray to C. Darwin, 4 Oct. 1862, *Life and Letters of Asa Gray*, 2:488–89, italics added.

69. Desmond and Moore, *Darwin*, 515–16.

70. Lurie, *Agassiz*, 265–66.

71. Lyell, *Antiquity*, 387–88 (see chap. 2, n. 37). Lyell also noted the analogous difficulties of defining *species* and *race* (388–89).

72. Charles Lyell, *Principles of Geology*, 12th ed. (London, 1875), 2:474–78, 480–81, quotation at 481.

73. Lyell, *Antiquity*, 446, 448–53.

74. Ibid., 454.

75. Liba Taub notes that Lyell's analogic reasoning at this point conveyed his "qualified support" of Darwin's theory. Taub, "Evolutionary Ideas," 171, 175 (see Prologue, n. 4).

76. Martin Rudwick, introduction to Lyell, *Principles*, 1:xviii–xix (see Prologue, n. 10).

77. Lyell, *Antiquity*, 454. Significantly, Lyell continued to cite Müller's "Comparative Mythology" of 1856, which had appeared prior to Darwin's *Origin*, no doubt because Müller's more recent *Lectures on the Science of Language* (vol. 1, 1862) was in part a critique of Darwin's theory.

78. Lyell, *Antiquity*, 457.

79. Ibid.

80. Ibid., 458.

81. Ibid., 458–62.

82. Ibid., 462. Lyell "recognized that the very same objections which might be brought against Darwin's theory could also be brought against a theory of linguistic descent, yet few would argue against the latter." Taub, "Evolutionary Ideas," 174.

83. The exception was Darwin's analogy of genealogical classification in chap. 13 of the *Origin*, which Lyell abandoned. Lyell would have seen that Darwin's ethnological analogy, conflating race in language, implied an obsolete and relatively short human chronology.

84. Lyell, *Antiquity*, 463–64.

85. Ibid., 467.

86. Ibid. See *Lyell's Scientific Journals*, 352.

87. Lyell, *Antiquity*, 469. Joachim Gessinger suggests that Lyell's analogy breaks down when he touches on grammar rather than word morphology. Actually, by noting the apparently predetermined patterns of grammar, Lyell was intentionally elaborating his bid to point out the explanatory limitations of Darwinism. See Gessinger, "Charles Lyell und Charles Darwin: Aktualismus und Evolution in der Geschichte der Sprachen," in *Language and Earth: Elective Affinities between the Emerging Sciences of Linguistics and Geology*, ed. Bernd Naumann, Frans Plank, and Gottfried Hofbauer (Philadelphia: John Benjamins, 1992), 323–56.

88. Lyell, *Antiquity*, 469. Here Lyell quoted Wilhelm von Humboldt's saying that "man is man only by means of speech, but in order to invent speech he must be already man." Max Müller, in many ways a disciple of Humboldt, had quoted this statement: *Lectures on the Science of Language*, 1:354.

89. Richard Whately, *Introductory Lectures on Political Economy* (1831; reprint, London, 1855), 59–65. Whately contended that political economy's picture of man in society offered a superior vindication of the divine wisdom and was hence the most striking version of natural theology. The Harvard moral philosopher Francis Bowen repeated Whately's argument, including the examples of the London economy and the beehive, in his *Principles of Political Economy* (Boston, 1856), 20–22, 27.

90. Baden Powell, *The Unity of Worlds and Nature: Three Essays*, 2d ed. (London, 1856), 90, 142–49. On Powell's natural theology, see Mary Pickard Winsor, "Louis Agassiz and the Species Question," *Studies in History of Biology* 3 (1979): 103–4.

91. See Frederick R. Prete, "The Conundrum of the Honey Bees: One Impediment to the Publication of Darwin's Theory," *Journal of the History of Biology* 23 (1990): 271–90. In the 1850s, Darwin had complained that honeybees were "potentially fatal" to his theory in part

because of their hive-building instinct. "This instinct was considered a true wonder of nature both by natural theologians and by the more secularly minded of the time. By what means the honey bee could have become endowed with the precise geometer's knowledge of how to construct the perfect hexagonal cell had been discussed extensively for centuries and was an often-used argument for the existence of God" (274–75).

92. Here Lyell's analogizing reflected an eclectic blend of design arguments, a theme he confirmed in the final pages of *Antiquity*. This was a change for Lyell, and one that probably reflected Asa Gray's influence. Initially, Lyell and Gray had approached the problem of design in Darwinism from opposite directions. In his journal, Lyell had regarded design as the progressive manifestation of divine intelligence. Gray, however, rejected this idealist brand of natural theology and regarded "adaptation to purpose" as the sole proof of design. After reading Gray's *Atlantic* series, Lyell professed to embrace this view as well: he affirmed that "the perpetual adaptation of the organic world to new conditions leaves the argument in favor of design, and therefore of a designer, as valid as ever." Lyell, *Antiquity*, 506. Lyell (502, 505) thereby adopted Gray's more flexible view of evolutionary causation: "Consistently with the derivative hypothesis, we may hold any of the popular views respecting the manner in which the changes of the natural world are brought about."

93. Lyell, *Antiquity*, 469.

94. Ibid., 468. See also *Lyell's Scientific Journals*, 422.

95. C. Darwin to Asa Gray, 23 Feb. 1863, Asa Gray Collection, Archives of the Gray Herbarium, Harvard University, italics in original. C. Darwin to J. D. Hooker, 24 Feb. 1863, *Life and Letters of Charles Darwin*, 3:7–8. Hooker replied: "I have finished Lyell and am enchanted with the Glacial Chapters, language, and the whole treatment of the Origin and Development subjects (with above qualifications)." J. D. Hooker to C. Darwin, 15 Mar. 1863, *Life and Letters of Sir Joseph Dalton Hooker*, ed. Leonard Huxley, 2 vols. (London, 1918), 2:34. C. Darwin to C. Lyell, 6 Mar. [1863], Lyell Collection, American Philosophical Society (289), Philadelphia.

96. Lyell, *Antiquity*, 469.

97. Lyell himself likewise defended his detached stance on strategic grounds: "But you ought to be satisfied, as I shall bring hundreds toward you, who if I treated the matter more dogmatically would have rebelled." C. Lyell to C. Darwin, 11 Mar. 1863, *Life, Letters and Journals of Sir Charles Lyell*, 2:363–64.

98. Asa Gray to C. Darwin, 20 Apr. 1863, *Life and Letters of Asa Gray*, 2:503–4.

99. C. Darwin to Asa Gray, 11 May [1863], Asa Gray Collection, Archives of Gray Herbarium, Harvard University. C. Darwin to J. D. Hooker, [9 May 1863], Darwin Collection, DAR 115:192, Cambridge University Library. When Darwin said that the notion of each small variation not being designed was a point that Lyell had "shirked," even though Darwin had "urged him to grapple" with it, he probably referred not to Lyell's use of the linguistic analogy per se but to the problem itself—that of believing that "each trifling variation" was designed.

100. See "Max Müller's Equivocal Role," above. Darwin used this phrasing consistently. He had urged Lyell, for example, to "by all means read Preface in about 20 pages, of Hensleigh Wedgwood's new Dictionary on first origin of Language." C. Darwin to C. Lyell, 10 Jan. [1860], *Correspondence of Charles Darwin*, 8:28.

101. Asa Gray to C. Darwin, 20 Apr. 1863, *Life and Letters of Asa Gray*, 2:503–4. Asa Gray to C. Darwin, 4 Oct. 1862, ibid., 488–89. Earlier, Gray had added a hopeful note: "I think I see

indications of a way out." Presumably, he referred to the kind of argument he had made in defense of design in the *Atlantic* series: in the sphere of language, if small changes were blindly fortuitous and unintended, one could still point to whole languages as manifesting design. Yet he dropped this hopeful theme after reading Lyell and thus probably saw the problem as more serious by then.

102. Gray, "Natural Selection," 121–22. Philosophers such as William Paley and William Smellie had tried to address this problem of the unbalanced ratio between resources and population, the manifest cruelty of Malthus's nature. They argued that this destructive superabundance of animal life at least produced some "utilities." See, for instance, William Smellie, *The Philosophy of Natural History* (Boston, 1845), 210–27.

103. Robert M. Young, "Malthus and the Evolutionists," *Past and Present* 43 (1969): 109–45, esp. 129.

104. "Antiquity of Man," *Athenaeum* (14 Feb. 1863): 221; J. D. Forbes, "Lyell's Antiquity of Man," *Edinburgh Review,* July 1863, 295; John Phillips, "Antiquity of Man," *Quarterly Review (London)* 114 (Oct. 1863): 414; C. H. Hitchcock, "The Antiquity of Man," *North American Review* 97 (Oct. 1863): 474. Surprisingly, an article-length review of Lyell's book by Hensleigh Wedgwood's daughter said nothing about the linguistic chapter: [Frances Julia Wedgwood], "Sir Charles Lyell on the Antiquity of Man," *Macmillan's Magazine* 7 (Apr. 1863): 476–87.

105. D. R. Goodwin, "The Antiquity of Man," *American Presbyterian and Theological Review,* n.s., 2 (1864): 257–58; "Antiquity of Man," *Spectator* (London), 7 Mar. 1863, 1725–1727. D. R. Goodwin made this point even in the midst of quoting from Lyell: " 'In our attempts to account for the origin of species'—we heartily thank Sir Charles Lyell for this clear and pointed statement,—'we find ourselves still sooner brought face to face with the working of a law of development of so high an order.' " Goodwin, "Antiquity of Man," 254.

106. W. Whewell to C. Lyell, 28 Feb., 14 May, 5 June, 1863, quoted in I. Todhunter, *William Whewell, D.D.: An Account of his Writings* (1876; reprint, New York: Johnson Reprint Corp., 1970), 429–32. Whewell admitted that he considered past changes in language to have been of a "superior order" to those in operation at present: "I do so, because my masters in the science of language, Pritchard and the like, do so" (431). Yet this appeal hurt Whewell's case, for J. C. Pritchard's writings came prior to, and so did not reflect, the human time revolution.

107. T. H. Huxley, "On the Methods and Results of Ethnology," *Fortnightly Review* 1 (1865): 257–77, reprinted in Huxley, *Critiques and Addresses* (New York, 1873): quotations at 218–19, 215. T. H. Huxley to F. Max Müller, 15 June 1865, quoted in Mario A. di Gregorio, *T. H. Huxley's Place in Natural Science* (New Haven: Yale University Press, 1984), 380–81; see further discussion, 160–61. For the same case made later, see A. H. Sayce, "Language and Race," *Journal of the Anthropological Institute of Great Britain and Ireland* 5 (1876): 212–17.

108. See Trautmann, "Revolution"; Trautmann, *Morgan,* 205–30 (see Prologue, n. 4).

109. Lyell, *Antiquity,* 386, quoted in Stocking, *Victorian Anthropology,* 75–76.

110. In pursuing the question of the common origin of all languages, "it is to be presumed that the classification of language is not, for obvious reasons, and as Darwin would contend, to be derived from the classification of race." The biblical genealogies were therefore framed in terms of "*blood,* not language." "The Lectures and Essays of Max Müller," *Christian Remembrancer* 56 (1868): 155, italics in original.

111. Review of "Origin of Human Races," by A. R. Wallace, *The Natural History Review,*

n.s., 4 (1864): 329, italics in original. The Darwinian affinity for monogenesis could also be interpreted as a matter of biological reality: Wallace, for instance, argued in 1864 that racial diversification itself could be explained in terms of natural selection. A. R. Wallace, "The Origin of Human Races and the Antiquity of Man deduced from the theory of 'Natural Selection,'" *Transactions of the Anthropological Society of London* 1 (1864): clviii–clxxxvii.

An example of the continued racial-monogenesis analogy appears in W. B. Carpenter's study of the genus Formaninifera: "Now the modifications which any single type of Formaninifera must have undergone, to give origin to the whole series of diversified forms now presented by that group, *are not greater in comparison with the modifications of which we have direct evidence,* than are those which the advocate for the specific unity of the human races has no hesitation in assuming as the probable account of their present divergence." Carpenter, "General Results of the Study of Typical Forms of Formaminifera, [etc.]," *Natural History Review,* n.s., 5 (April 1865): 196, italics in original.

112. Lyell kept the protransmutationist analogy alive in the revised tenth edition (1868) of his *Principles of Geology.* While addressing the species question, Lyell as usual rejected Louis Agassiz's argument by analogy that the "parent stocks" of Europeans, Negroes, and other main human types were originally distinct. In this context, he added the comparison with language: "That it should be so difficult for us to picture to ourselves the manner in which a species may be elaborated by Variation and Selection, need not surprise us when we consider how hard it is to obtain a clear idea of the growth and establishment of a new language, even when we are sure that the same has originated only a few centuries before our time." It would be hard, for example, to "fix upon the exact year or generation" when the English language was formed or to mark off clearly its transitional phases (475–76).

And yet, Lyell implied, philologists were still able to infer that languages had so evolved. Here again was the theme of the ultimate success of philology in spite of the inherent limitations under which it was obliged to work. In good Lyellian (and Darwinian) fashion, Lyell here argued *via negativa,* for a removal of obstacles to Darwin's theory. The obstacle in view was the insufficiency of evidence demonstrating a gradual emergence from something that had existed earlier. Lyell thus made the comparison in the old modest way: his goal was not actually to argue that transmutation had occurred; rather, it was to get readers to acknowledge that something "difficult for us to picture" could be known, nonetheless, to be true.

Chapter 4. The Convoluted Path to The Descent of Man

1. A second translation, in 1877, was done by Viktor Carus.

2. August Schleicher, *Die Darwinsche Theorie und die Sprachwissenschaft. Offenes Sendschreiben an Herrn Dr. Ernst Haeckel* (Weimar, 1863).

3. August Schleicher, *Darwinism Tested by the Science of Language,* trans. Alexander V. W. Bikkers (London, 1869), 14–16, reprinted (with identical pagination) in *Linguistics and Evolutionary Theory: Three Essays by August Schleicher, Ernst Haeckel, and Wilhelm Bleek,* ed. Konrad Koerner (Philadelphia: John Benjamins, 1983). This English translation of Schleicher's pamphlet substantiated the claim by appending a relevant passage, translated, from Schleicher's *Die Deutsche Sprache* (1860).

4. Schleicher, *Darwinism Tested,* 17–18.

5. Ibid., 17. Schleicher said that his aim was "to point out how the main features of Darwin's theory are applicable to the life of languages, *or even, we might say, how the development of human speech has already been unconsciously illustrative of the same.*" Ibid., 30, italics in original.

6. Ibid., 64.

7. Ibid., 35.

8. Ibid., 40–41.

9. Ibid., 43.

10. Ibid., 45. A long section (45–64) deals with the question of the ultimate origin of linguistic families.

11. Ibid., 66.

12. At the earliest stages of a language group's history, this conclusion really could be reached only indirectly, via the comparative method of reconstruction. Yet Schleicher did not allude to the reconstructive technique that he had done so much to pioneer. He focused, rather, on the hard textual evidence from which linguistic genealogies were extrapolated. It was this evidentiary record that was more complete for philology than for paleobiology.

13. Lyell, *Antiquity*, 457 (see chap. 2, n. 37).

14. Desmond and Moore describe the *Reader* as "probably the last attempt in Victorian England to keep together liberal scientists, theologians, and men of letters." Desmond and Moore, *Darwin*, 527 (see chap. 2, n. 15).

15. "The Darwinian Theory in Philology," *Reader* 3 (27 Feb. 1864): 261–62.

16. Ibid.

17. Thomas H. Huxley, "Criticisms on 'The Origin of Species'," in his *Darwiniana: Essays* (New York, 1896), 80–82. This article originally appeared in *Natural History Review*, n.s., 4 (1864): 566–80. Desmond and Moore, *Darwin*, 526–27.

18. John F. Byrne, "*The Reader*: A Review of Literature, Science and the Arts, 1863–1867" (Ph.D. diss., Northwestern University, 1964); David A. Roos, "The 'Aims and Intentions' of *Nature*," in *Victorian Science and Victorian Values: Literary Perspectives*, ed. J. Paradis and T. Postlewait (New Brunswick, N.J.: Rutgers University Press, 1985), 163; *Life and Letters of Thomas Henry Huxley*, ed. Leonard Huxley, 2 vols. (London, 1900), 1:211.

19. F.W.F., "The Public Schools Report," *Reader* 3 (11 June 1864): 738–39. *Life and Letters of Thomas Henry Huxley*, 1:277, 308. Farrar later told Darwin that he was doing all he could to "encourage a taste for observation and for Natural History" among his charges at Marlborough College. "It is my very strong desire to do more for education in science as opportunity offers." F. W. Farrar to C. Darwin, 21 Feb. [1871], Darwin Collection, DAR 164, Cambridge University Library.

20. *Dictionary of National Biography*, ed. Leslie Stephen and Sidney Smith (New York: Macmillan, 1908–9), s.v. "Farrar, Frederic William"; F. W. Farrar to C. Darwin, 1 and 5 Feb. 1866; C. Darwin to F. W. Farrar, 3 Feb. 1866, summarized in *A Calendar of the Correspondence of Charles Darwin, 1821–1882*, ed. F. Burkhardt (New York: Cambridge University Press, 1994).

21. [F. J. Wedgwood], "Origin of Language," 54, italics in original (see chap. 3, n. 55); Müller, *Lectures on the Science of Language*, 2:15 (see Prologue, n. 8).

22. Farrar's linguistic theory was well appreciated by the utilitarian and secularist *Westminster Review*. "Origin of Language," *Westminster Review* 86 (1866): 101. Farrar and

Wedgwood were not the only philologists to critique Max Müller on the issue of the origin of language. Others included F. J. Furnivall, W. D. Whitney, and Bertold Delbrück. See, for example: [F. J. Furnivall], "Language Books New and Old," *Reader* 6 (14 Oct. 1865): 420–21.

23. F. W. Farrar, preface to his *Essay on the Origin of Language* (London, 1860).

24. C. Darwin to Asa Gray, 6 Nov. 1862, *Life and Letters of Charles Darwin*, 2:182 (see chap. 2, n. 61).

25. Müller, *Lectures on the Science of Language*, 1:42, 44, 49. Müller's (and Schleicher's) view of human agency in regard to linguistic change was an extreme version of the impulse to separate essential linguistic development from the accidents of history, the conquests, culture contact, and other ways in which languages changed through the free action of historically conscious actors. Stam, *Origin of Language*, 238 (see chap. 1, n. 6).

26. F. Max Müller, "Lectures on Mr. Darwin's Philosophy of Language," *Fraser's Magazine*, n.s., 7 (May-June 1873); F. Max Müller to C. Darwin, 7 Jan. 1875, *The Life and Letters of the Right Honourable Friedrich Max Müller; edited by his wife*, 2 vols. (London, 1902), 1:476.

27. Müller, *Lectures on the Science of Language*, 2:326–27. At least one reviewer noted favorably Müller's analogies with Darwinism: review, "Müller's Lectures on the Science of Language, Second Series," *Daily News (London)*, 21 Oct. 1864.

28. [F. M. Müller], "Comparative Philology," *Edinburgh Review* 94 (Oct. 1851): 300–303, 312, 314. Here Müller only hinted at his later view of dialects as existing prior to unified languages such as Latin.

29. Müller's disciple George Wilson Cox confirmed that Müller's dialects theory was a new departure since his 1856 Oxford lecture. G. W. Cox, review of *Lectures on the Science of Language*, vol. 2, by Müller, *Westminster Review* 83 (Jan. 1865): 35–64.

30. Müller, *Lectures on the Science of Language*, 1:60, 194–95. At this time, there was widespread interest in dialect production as a concrete aspect of ethnology. See Report on Meeting of the Ethnological Society, and Roderick I. Murchison, letter, "Diversity of Dialects," *Reader* 3 (13 Feb. 1864): 249; and 3 (20 Feb. 1864): 241.

31. Müller, *Lectures on the Science of Language*, 196–97. Müller explained: "Many of the words which give to French and Italian their classical appearance, are really of much later date, and were imported into them by medieval scholars, lawyers and divines." Ibid. See also F. Max Müller, "Introductory Lecture on the Science of Language," *Macmillan's Magazine* 7 (Mar. 1863): 346, 348. Max Müller, "The Science of Language," *Saturday Review* 24 (Nov. 30, 1867): 700.

32. Review, "Professor Max Müller's Lectures on Language: Second Series," *Reader* 4 (20 Aug. 1864): 220–21. The reviewer was no doubt Farrar: he was well acquainted with and generally complimentary of Müller and his work, knew German, and stressed the idea of philology as elevated to a "science."

33. Frederic W. Farrar, *Chapters on Language* (London, 1865), 49–51.

34. C. Darwin to F. W. Farrar, 2 Nov. 1865, *The Life of Frederic William Farrar*, ed. Reginald Farrar (New York, 1904), 107–8.

35. Ibid.

36. F. W. Farrar to C. Darwin, 6 Nov. 1865, Darwin Collection, DAR 164:35, Cambridge University Library, italics in original. Farrar later confused this letter with his reply to Darwin on receiving a copy of *The Descent of Man* in 1871. His account of this exchange of letters, published in the *Life* compiled by his son, is nonetheless revealing. Here Farrar recalls explaining his polygenist reservations to Darwin, especially as supported by physical

anthropology's evidence of ancient racial distinctions. Farrar then reports a characteristic, albeit unattested, reply from Darwin. "Mr. Darwin admitted the fact, but made this very striking answer: '*You are arguing from the last page of a volume of many thousands of pages.*'" *Life of Frederic William Farrar*, 108. The memoir in which Farrar wrote this account was apparently in the possession of his family and never published.

37. F. W. Farrar to C. Darwin, 6 Nov. 1865, Darwin Collection, DAR 164:35, Cambridge University Library, italics in original.

38. The extreme difference between isolating and inflecting could be seen by a comparison of Chinese and French. The Chinese word for *twenty*, for example, was a simple juxtaposition of sounds meaning "two-ten." The French *vingt*, by contrast, melded early versions of *deux* and *dix* without preserving their distinctness.

39. F. Max Müller, *On the Stratification of Language* (London, 1868), 13, 18.

40. F. W. Farrar, "The Growth and Development of Language," *Journal of Philology* (*Cambridge*) 1 (1868): 20–21, italics in original. The same argument, again not specifically alluding to Darwinism, was made by an American theologian who was skeptical of the developmentalist view of the three linguistic morphological types: "These groups of languages lie well defined and in classification widely separated from one another; if the theory were true, should we not find the intermediate spaces filled with languages here just emerging from one state, there just preparing for transition into another?" [C. A. Aiken], "Whitney on Language," *Princeton Review*, Apr. 1868, 283.

41. Farrar added a postscript to his published "Growth and Development," declaring that he had written his lecture prior to receiving a copy of Max Müller's "On the Stratification of Language." Farrar, "Growth and Development of Language," 23. His argument nonetheless made a perfect answer to Müller. Farrar repeated that argument a short time later, in an actual review of Müller's *Stratification*. F. W. Farrar, review of *On the Stratification of Language*, by F. Max Müller, *Fortnightly Review*, n.s., 4 (1868): 346–48. This review was generally unfavorable.

42. F. W. Farrar, "Philology as One of the Sciences," *Macmillan's Magazine* 19 (1869): 256–57. Müller's schema provides the outline for much of his *Lectures on the Science of Language*, vol. 1.

43. Farrar, "Philology as One of the Sciences," 256–57.

44. Ibid., 257.

45. The translator, Alexander V. W. Bikkers, was a Dutch scholar of Oriental languages who lived in England.

46. Review, *Darwinism Tested by the Science of Language*, *Athenaeum* (27 Nov. 1869, 699). This review was done in a popular style, different from that of any of the writers we have seen in this chapter.

47. W. S. Dallas, "Science: Review of Contemporary Literature," *Westminster Review* n.s., 37 (Jan.-Apr. 1870): 288.

48. Roy M. Macleod, "Is It Safe to Look Back?" *Nature* 224 (1969): 445. *Nature* was originally published by the house of Macmillan under the long-running editorship of the astronomer J. H. Lockyer. It superseded the *Reader* and the *Natural History Review*.

49. Wholly apart from Darwinian considerations, Müller aimed a good half of this review across the ocean, albeit tacitly, at his American nemesis, William Dwight Whitney (1827–94).

50. F. Max Müller, "The Science of Language," *Nature* 1 (6 Jan. 1870): 257, 258. Müller did

make a sensible criticism of Schleicher's disembodied view of linguistic dominance and extinction, noting that languages do not die out but the peoples who speak them do. Rather than emphasize the struggle for life among separate languages, the better analogy was "the struggle for life among words and grammatical forms which is constantly going on in each language. Here the better, the shorter, the easier forms are constantly gaining the upper hand" (258).

51. Müller, "Science of Language" (*Nature*), 258.

52. Ibid. Here Müller also, of course, abandoned the antievolutionism of his 1861 lecture.

53. Admittedly, this distinction was somewhat unclear in the relevant discussion of natural selection, divergence, and extinction in chap. 4 of the *Origin*. See Darwin, *Origin*, 116–28 (see chap. 2, n. 27). See also Bowler, *Evolution*, 160–62, 176, 179–81 (see Prologue, n. 12), on the difference between interspecies competition for territory (a phenomenon noted prior to Darwin's *Origin* by Lyell and others) and Darwin's theory of individual competition. Only the latter led to natural selection and transmutation, the "struggle for existence" among species led chiefly to the divergence of neighboring species' character traits so as to fill the available environmental niches.

54. Müller possibly meant to identify analogically Schleicher's Indo-European protolanguage with Richard Owen's ideal vertebrate archetype. This, he must have known, would hurt Schleicher's reputation in Darwin's eyes.

55. F. W. Farrar, "Philology and Darwinism," *Nature* 1 (24 Mar. 1870): 527. Farrar said that he had actually written this piece half a year earlier, before Schleicher's pamphlet appeared in translation. Such a study of Schleicher's argument would have been a natural part of Farrar's preparation for his "Families of Speech" lectures of 1869. Yet the printed review of Schleicher worked perfectly as still another rebuttal of Müller, and Farrar's claim to have written the piece prior to the appearance of Müller's review in *Nature* strains Farrar's credibility. See also n. 41 above.

56. Farrar, "Philology and Darwinism," 528.

57. Although Farrar was a racial polygenist, this stance on ultimate ethnic origins did not at all preclude him from affirming the branching pattern in the more localized Indo-European family. Hence there was no contradiction in his use of the analogy on this point.

58. Farrar, "Philology and Darwinism," 528.

59. Ibid. A similar case of Farrar's suspended judgment concerning scientific issues, c. 1870, is described in Emel Aileen Gökyigit, "The Reception of Francis Galton's *Hereditary Genius* in the Victorian Periodical Press," *Journal of the History of Biology* 27 (1994): 229–36. Gökyigit classes Farrar among the "neutral" reviewers of Galton's argument.

60. J. Vernon Jensen, "The X-Club: Fraternity of Victorian Scientists," *British Journal for the History of Science* 5 (1970): 63–72.

61. Frederic W. Farrar, *Families of Speech* (London, 1870), 52–53. (Farrar dedicated this work to Max Müller.) In this passage, Farrar borrowed directly from Frances Julia Wedgwood's 1862 article urging the use of a Lyellian methodology in linguistics. "Our science occupies, at this day, the position of geology forty years ago. Those among us who can look so far back may remember the smile of derision with which we heard that Scrope and Lyell were accounting for the formation of continents and elevation of mountains by the mere continuance of those agencies which we see working at the present day in the crumbling of our sea-cliffs, the sediment of our rivers." See [F. J. Wedgwood], "Origin of Language," 59.

62. Farrar, *Families of Speech*, 55–56.

63. Ibid., 56. This reciprocal analogy appeared as well in (Farrar, "Philology and Darwinism," 529). Here Farrar also made one of the very few references to Darwin's linguistic analogizing in *The Origin of Species*: Darwin, he said, "devotes a few words to the classification of languages as affording a confirmation of his theories"—referring clearly to the ethnological analogy in chap. 13. Farrar added that linguistic classification conveyed this confirmation "to an extent of which probably he [Darwin] was not at first aware." Farrar referred especially to "the immense changes which can be effected by infinitesimally gradual modification," unaware that Darwin had made this same point in the unpublished *bishop* analogy.

64. Müller, *Lectures on the Science of Language*, 2:15.

65. Review of *Families of Speech*, by Frederic W. Farrar, *Athenaeum* (5 Feb. 1870): 189. Müller's authorship of this review is suggested also by its attack on the motion of language-based ethnological genealogy.

66. Desmond and Moore, *Darwin*, 527.

67. Ibid., 478–79.

68. Nicolaas A. Rupke, *Richard Owen, Victorian Naturalist* (New Haven: Yale University Press, 1994), 170–71, 236–37, 209–10, 238–40, 204–5.

69. George Douglas Campbell, duke of Argyll, *The Reign of Law* (London, 1866), 213–14, 31–32, 75–76.

70. Ibid., 241–60.

71. Ibid., 232; Desmond and Moore, *Darwin*, 545–46.

72. Darwin, *Variation*, 2:371–72 (see chap. 2, n. 26). Bowler glosses Darwin's argument: "If variations were directed along beneficial lines, selection would become superfluous; the variations themselves would direct the course of evolution. In any case, we observe that the majority of variations are meaningless." Bowler, *Evolution*, 224.

73. The passage from *Anatomy of Vertebrates*, vol. 3 (1868), Owen's fullest statement on evolution, also came out as a forty-page booklet, *Derivative Hypothesis of Life and Species* (1868). Rupke, *Owen*, 248–50; Desmond and Moore, *Darwin*, 527, 534, 543–46. Asa Gray told Darwin that Argylle's views were similar to his own: "his main points are those I hammered out in the 'Atlantic' etc; indeed I see signs of his having read the same." Asa Gray to C. Darwin, 22 Mar. 1863, *Life and Letters of Asa Gray*, 501–2 (see chap. 3, n. 39).

74. Desmond and Moore, *Darwin*, 572.

75. Charles Darwin, *The Descent of Man and Selection in Relation to Sex* (1871; reprint, Princeton: Princeton University Press, 1981), 188–89. Unless otherwise stated, further references are to this edition.

76. Ibid., 53. Here Darwin referred to Hensleigh Wedgwood's *Origin of Language* (1866).

77. Darwin, *Descent*, 58. A useful discussion of the passage about the origin of language in *Descent* appears in Stam, *Origin of Language*, 245–46.

78. Darwin, *Descent*, 59. The revised wording appears in Darwin, *The Descent of Man and Selection in Relation to Sex*, rev. ed., vol. 21 of *Works of Charles Darwin*, pt. 1, 94. The paragraph containing the language-species analogy remained otherwise unchanged in 1874 and in the final "revised and augmented" edition of 1877.

79. Darwin, *Descent* (1st ed. [1871], and so hereafter), 60. See Darwin, *Origin*, 114–15.

80. Darwin, *Descent*, 60; Darwin, *Origin*, 106; Darwin, *Descent*, 60.

81. Darwin alluded to Lyell, *Antiquity*, 362; Darwin, quoted in Farrar, "Philology and Darwinism," 528.

82. Darwin, *Descent*, 60. Darwin, quoted in Müller, "Science of Language," 257.

83. Darwin, *Descent*, 60.

84. There is no direct evidence that Darwin owned or read Farrar's *Families of Speech*.

85. Beer, *Darwin's Plots*, 54–55 (see Prologue, n. 1), suggests this thesis, that by portraying language as "authenticated by the natural order," Darwin rendered it an especially apt source of analogy representing that order. Beer also notes the loss of "analogical function" in the passage in *Descent*, along with the resulting "self-proving argument" and "self-verifying interchange" of images. She regards the *Descent* passage as an extreme terminus in the reciprocal transfer of metaphors between linguistics and biology, a point at which each field was forced to turn to explanatory models intrinsic to its subject matter. Beer, "Darwin and the Growth of Language Theory," 164–65 (see Prologue, n. 4). This perceptive interpretation slights, however, the consciously polemical intent with which Darwin constructed his *Descent* analogies.

86. An example was Darwin's opportunistic use of Max Müller's analogizing. Another example appears in *Descent*, rev. ed., 102 n. 53, in which Darwin seized upon one of the rare instances in which the philologist W. D. Whitney acknowledged that linguistic change was unintended and "unconscious."

87. Darwin, *Descent*, 233.

88. Ibid., 228, 229. Darwin added that the term *species* was not the main issue, since there was no agreement as to its definition (228–29).

89. This was the very thing Charles Lyell had pointed out on several occasions during the 1860s, when addressing Louis Agassiz's racial polygenism.

90. Trautmann, "Revolution, 379–97 (see chap. 3, n. 11).

91. Ibid., 75–76. Users of the linguistic analogy, especially those who were anti-Darwinians or theistic evolutionists, never could agree among themselves whether the unconscious aspect of linguistic change worked for or against their positions.

92. Schleicher, *Darwinism Tested*, 20–21.

93. William Taylor, letter to *Nature* 2 (19 May 1870): 48; Arthur Ransom, letter to ibid., 2 (9 June 1870): 103–4. The *Wellesley Index to Victorian Periodicals* lists Ransom as a "miscellaneous writer." Another correspondent responded to Farrar's piece on Schleicher by noting a weakness in the analogy, that "untamed" dialects, such as those in Scotland, were more variable than the Queen's English, whereas "tamed" species, under domestication, showed greater variability than those in the wild. S.J., letters to *Nature* 2 (26 May 1870): 66, and 3 (17 Nov. 1870): 48.

94. William Taylor, letter to *Nature* 2 (29 Sept. 1870): 435.

95. William Taylor, "The Variation of Languages and Species," *British and Foreign Evangelical Review*, Oct. 1871, 719.

Chapter 5. A Convergence of "Scientific" Disciplines

1. Miles, *Student of Philology*, 47–49.

2. See the essays in the first half of Hoenigswald and Wiener, eds., *Biological Metaphor and Cladistic Classification* (see chap. 2, n. 12).

3. Surveys of generic evolutionism in the various disciplines include Brooklyn Ethical Association, *Evolution in Science, Philosophy, and Art* (New York, 1891); Frederick O. Bower, *Evolution in the Light of Modern Knowledge: A Collective Work* (London, 1925).

4. Note that this juxtaposition of diagrams did not appear in an English version until 1910. A useful overview of Haeckel's books and tree diagrams appears in Jane M. Oppenheimer, "Haeckel's Variations on Darwin," in Hoenigswald and Wiener, *Biological Metaphor and Cladistic Classification*, 123–35.

5. Ernst Haeckel, *History of Creation*, trans., E. Ray Lankester, vol. 2 (New York, 1883), 301–3; Haeckel, *The Evolution of Man*, vol. 2 (New York, 1910), 18–27, quotations at 24.

6. Ibid., 27.

7. For example, Georg Uschmann, *Ernst Haeckel: Biographie in Briefen* (Gütersloh: Prisma Verlag, 1984), 10. See the long passage in Schleicher, *Darwinism Tested*, 21–30 (see chap. 4, n. 3), where he describes the character of modern science as philosophically monist.

8. Konrad Koerner, "Schleichers Einfluss auf Haeckel: Schlaglichter auf die Abhängigkiet zwischen linguistischen und biologischen Theorien im 19. Jahrhundert," in Koerner, *Practicing Linguistic Historiography: Selected Essays* (Philadelphia: John Benjamins, 1989), 211–14; Uschmann, *Haeckel*, 101.

9. Koerner, "Schleichers Einfluss auf Haeckel," 211–14; Schleicher, *Darwinism Tested*, 19–20. The common mistake of supposing Darwinian influence on Schleicher's linguistics appears, for example, in Colin Renfrew, *Archaeology and Language: The Puzzle of Indo-European Origins* (London: Jonathan Cape, 1987), 102.

10. On Gegenbaur, see William Coleman, "Morphology between Type Concept and Descent Theory," *History of Medicine and Allied Sciences* 31 (1976): 149–75. Contra my argument, Bowler implies that Gegenbaur influenced Haeckel in this area. Bowler *Evolution*, 200–202 (see Prologue, n. 12).

11. E. S. Russell, *Form and Function: A Contribution to the History of Animal Morphology* (reprint, 1916; Chicago: University of Chicago Press, 1982), 250–51, 260; Di Gregorio, *Huxley's Place*, 77–80 (see chap. 3, n. 107).

12. G. J. Romanes, *Darwin, and After Darwin*, 2 vols. (1892; reprint, Chicago, 1910), 1:32.

13. Farrar had taken that thesis to its most extreme, exaggerating the sureness of comparative philology's conclusions. He had declared that the *Origin*'s pedigree diagram was "to a great extent ideal and hypothetical; while the [comparative philologist's] table of languages is merely an expression of indisputable discoveries." Farrar, "Philology and Darwinism," 528 (see chap. 4, n. 55). Here Farrar spoke truly enough, for he referred to the indisputable principle that Greek, Latin, and Sanskrit, as well as their descendants, were related to some common ancestor. Yet when it came to actually reconstructing that ancestor and the intermediate lineages among the branches of Indo-European, that work involved hypothesis.

14. Romanes, *Darwin*, 33; W. B. Scott, *The Theory of Evolution* (New York, 1917), 55–56.

15. Chauncey Wright, "The Evolution of Self-Consciousness," *North American Review* 116 (Apr. 1873): 301. Wright here defended the language-species analogy and challenged what he saw as W. D. Whitney's overemphasis on human agency. See William Dwight Whitney, "Schleicher and the Physical Theory of Language," in his *Oriental and Linguistic Studies*, 2 vols. (New York, 1873–74), 1:298–331.

16. John Fiske, "Whitney's *Language and the Study of Language*," *Nation* 5 (1867): 369. Fiske's view mirrored that of the work he was reviewing. W. D. Whitney had said that the analogy suggested itself quite naturally, although he did not draw it between the entire disciplines of philology and biology. Whitney had also commended Lyell's and Schleicher's

applications of the analogy to the Darwinian question. With characteristic caution, however, he said that the analogy could prove nothing about Darwinism itself. W. D. Whitney, *Language and the Study of Language* (New York, 1867), 46–47.

17. W. D. Whitney grew more skeptical with time. "That every successive phase of a historical institution is the outgrowth of a preceding phase, and differs little from it, is a truth long coming to clear recognition and fruitful application in every department of historic research, prior to and in complete independence of any doctrine of evolution in the natural world. Only error and confusion have come of the attempts made to connect Darwinism and philologic science." Whitney, review of *The Alphabet: An Account of the Origin and Development of Letters,* by Isaac Taylor, *Science* 2 (28 Sept. 1883): 439.

18. A. H. Sayce, *The Principles of Comparative Philology,* 2d ed. (London, 1875), xviii–xxi, 100–103, 125–31. Latter-day objections to the biological metaphor in linguistics are summed up in Robert D. Stevick, "The Biological Model and Historical Linguistics," *Language* 39 (1963): 159–69.

19. Abel Hovelacque, *The Science of Language: Linguistics, Philology, Etymology,* trans. A. H. Keane (London, 1877), 309–11. The original was A. Hovelacque, *La Linguistique* (Paris, 1876).

20. Hermann Paul, *Principles of the History of Language,* trans. H. A. Strong (London, 1888), 20–22 ff.

21. J. M. Edmonds, *An Introduction to Comparative Philology for Classical Students* (Cambridge, 1906), 82–84; Peter Giles, "Evolution and the Science of Language," in *Darwin and Modern Science,* ed. A. C. Seward (Cambridge, 1909), 526–28.

22. C. Darwin to E. Haeckel, 19 Nov. 1868, *Life and Letters of Charles Darwin,* 3:105 (see chap. 2, n. 61). Clearly inspired by Haeckel, Darwin drew up in private a phylogenetic tree diagram, showing the hypothetical relation of humanity to the other primates. Gruber, *Darwin on Man,* 197 (see chap. 2, n. 10).

23. T. H. Huxley, letter, "The Classification of Birds," *Ibis,* July 1868, reprinted in *The Scientific Memoirs of Thomas Henry Huxley,* ed. Michael Foster and E. Ray Lankester (London, 1901), 296; T. H. Huxley, "Ornithology," *Encyclopaedia Britannica,* 9th ed. (1881); Huxley, *Man's Place* 271 (see chap. 3, n. 66).

24. Bowler, *Evolution,* 202–5, 269–70. Biological research thereby supported the idea of ancestor/descendent relationships in the pattern of evolution while leaving open the explanation of the causal process underlying those relationships. The distinction between pattern and process hypotheses is set forth in Elliot Sober, *Reconstructing the Past: Parsimony, Evolution, and Inference* (Cambridge: MIT Press, 1988), 8–9.

25. Darwin, *Origin,* 116–30 (see chap. 2, n. 27).

26. C. Darwin to A. Gray, 11 May [1863], Asa Gray Collection, Archives of the Gray Herbarium, Harvard University, italics in original; C. Darwin, "Origin of Species," *Athenaeum* (9 May 1863): 617.

27. Mayr, *One Long Argument,* 26–27 (see Prologue, n. 11). Said one writer in the 1890s, in an encyclopedia article on zoology which made conspicuously little mention of natural selection: "The idea of descent gave for the first time a point from which all branches of the science could be viewed. The facts of systematic zoology, comparative anatomy, embryology, and the biological relations of animals could now be handled together for a common end, and their own inter-relations could now be seen." E. A. Birge, "Zoology," *Johnson's Universal Cyclopedia,* ed. C. K. Adams, 8 vols. (New York, 1894), 8:904.

28. The German-born Jacques Loeb presents an exception in this regard: he was the extreme case in America of dissent from the evolutionist thesis in biology, that thesis being the legatee of the old morphological emphasis on anatomical form, adaptation, and the geographic distribution of plants and animals. Loeb stressed physicochemical and environmental influences on the physiology of organisms and pushed to reorient biology from natural history to a kind of biological engineering. Philip J. Pauly, *Controlling Life: Jacques Loeb and the Engineering Idea in Biology* (New York: Oxford University Press, 1987), 79–83.

29. E. D. Cope, *The Origin of the Fittest: Essays on Evolution* (New York, 1887), 329; Cope, *The Primary Factors of Organic Evolution* (Chicago, 1896), 106, 110, 115; E. R. Lankester, "Zoology," *Encyclopaedia Britannica*, 11th ed. (1910–11); Henry Fairfield Osborn, *The Origin and Evolution of Life* (New York, 1918), 236; George Alfred Baitsell, ed., *The Evolution of Man* (New Haven, 1923), 36; J. B. S. Haldane and Julian Huxley, *Animal Biology* (Oxford: Clarendon Press, 1927), fig. 81; G. G. Simpson, *Tempo and Mode in Evolution* (New York: Columbia University Press, 1944), 213; Simpson, *The Major Features of Evolution* (New York: Columbia University Press, 1953), 261; William King Gregory, *Evolution Emerging* (New York: Macmillan, 1951), 175. Additional examples appear in Robin Craw, "Margins of Cladistics: Identity, Difference and Place in the Emergence of Phylogenetic Systematics, 1864–1975," in *The Tree of Life*, ed. Paul Griffiths (Boston: Klumer Academic Publications, 1992), 69–89; Patterson, "Contribution of Paleontology to Teleostean Phylogeny," 579–644 (see chap. 2, n. 13). In addition, the University of London zoologist Arthur Dendy gave perhaps the most powerful verbal exposition of the argument for the identity of the classificatory and phylogenetic trees of life. Dendy, *Outlines of Evolutionary Biology* (London, 1912), 240.

30. H. C. Chapman, *Evolution of Life* (Philadelphia, 1873), 69; F. W. Headley, *Problems of Evolution* (London, 1900), 138–41; Benjamin C. Gruenberg, *The Story of Evolution* (New York: D. Van Nostrand, 1929), 69–71.

31. Bowler argues that the branching tree schema continued to be infused with the older idea of goal-oriented, progressionist evolution from 1859 until the early twentieth century, at which time a truer appreciation of evolution's divergent character emerged: Bowler, "Darwinism and Modernism," 236–54, esp. 247–50 (see Prologue, n. 12). Haeckel, for example, sometimes portrayed a central trunk in the zoological tree running straight up to humankind (see fig. 5.3), consonant with his emphasis on embryology as the model of development among species. Bowler's cogent analysis of the difference between the late-nineteenth- and early-twentieth-century evolutionary trees should not, however, blind us to the forest they all occupied. Genealogy, albeit expressed in various family tree forms, served as the central evolutionary metaphor throughout this entire period.

32. The Neogrammarians were centered in German universities, yet their influence spread far. Starting in the 9th ed. (1881) of the *Encyclopaedia Britannica*, Eduard Sievers wrote the article "Philology, ii: Comparative Philology of the Indo-European Languages."

33. Only the comparative-historical approach in linguistics had been capable of producing such laws; typological classification of languages revealed no such historical correlations and posed few questions for productive research.

34. Henry Sweet, "Recent Investigations of the Indogermanic Vowel-System," *Transactions of the Philological Society* (1880–81): 155.

35. *International Encyclopedia of Linguistics*, ed. William Bright (New York: Oxford University Press, 1992), s.v. "Comparative Method." Wiener notes the obvious weakness of this technique. "Reconstructing protoforms from the evidence of daughter languages and then

constructing subgroupings on the basis of these protoforms may involve aspects of circularity." See Wiener, "Of Phonetics and Genetics: A Comparison of Classification in Linguistic and Organic Systems," in Hoenigswald and Wiener, *Biological Metaphor and Cladistic Classification*, 221.

36. W. F. Albright, "New Identifications of Ancient Towns," *Bulletin of the American Schools of Oriental Research* 9 (1923): 5–7.

37. C. Darwin, "The Doctrine of Heterogeny and Modification of Species," *Athenaeum* (25 Apr. 1863): 554; "Origin" (*Athenaeum*), 617; Darwin, *Notebooks*, 370 (Notebook D, 117).

38. Darwin, *Notebooks*, 355 (Notebook D, 67). See the discussion of this and the above passage in Ghiselin, *Triumph of Darwinian Method* (Berkeley: University of California Press, 1969), 235–36; and David L. Hull, *Darwin and His Critics* (Chicago: University of Chicago Press, 1973), 8, 9, 13.

39. Sayce, *Comparative Philology*, 6–13, 46.

40. Fritz Müller, *Facts and Arguments for Darwin* trans. W. S. Dallas (London, 1869), 1–2. A number of the historically oriented themes discussed in the following pages appear also in Stephen Jay Gould, "Evolution and the Triumph of Homology, or Why History Matters," *American Scientist* 74 (1986): 60–69. For instance, Gould notes the "nomothetic undertones" and the "postdiction" of past events in Darwin's scientific work as a whole, themes present in philology as well. My focus on branching genealogy, however, is narrower than Gould's discussion of Darwin's "historical science": the latter dealt not just with organisms but with geology and coral reefs as well, and so it was not confined to the branching schema. See Gould, "Evolution and Triumph of Homology"; Gould, "Darwinism and the Expansion of Evolutionary Theory," *Science* 216 (23 Apr. 1982): 380–87.

41. On the actualist/uniformitarian principle, Darwinism regarded the evolutionary process as perennial rather than exclusively past-oriented. Timelessness was manifest, for instance, in the *Origin*'s opening chapter on domestic breeding, for breeding was a contemporary practice meant to represent the workings of natural selection in the past, present, and future. This selection analogy, however, pertained to the Darwinian process of change, not its overall pattern. In overall pattern, both Darwin's theory and comparative philology dealt with the cumulative effects of completed and nonrepeatable events.

42. Sober, *Reconstructing the Past*, 6.

43. As Renfrew points out, the "limiting assumptions" held by nineteenth-century philologists "inevitably restricted the historical conclusions to which they came." Renfrew, *Archaeology and Language*, 99.

44. F. J. Teggart, *Theory of History* (New Haven: Yale University Press, 1925), 129, quoted in Margaret T. Hodgen, *The Doctrine of Survivals* (London: Allenson & Co., 1936), 184; Henry M. Hoenigswald, "On the History of the Comparative Method," *Anthropological Linguistics* 5 (1963): 2.

45. The cladistic revolution since midcentury (the German systematist Willi Hennig set forth the basic themes in the 1950s and 1960s) has reasserted the value of a phylogenetic classification of plants and animals. Evolutionists all, biological systematists have nonetheless disagreed on whether organisms may best be classified on cladistic grounds, on "phenetic" (typological) grounds, or on a conflation of these two which stresses adaptive traits, producing a so-called evolutionary classification. Defenders of cladistic method argue that the introduction of noncladistic traits confuses and compromises a strictly genealogical approach.

The impression is sometimes given that genealogical classification in biology declined in the early twentieth century and was revived only after midcentury by the cladistic revolution. Although this is true in technical systematics, for present purposes it is misleading, for there was continuity all along as major biologists used genealogical diagrams to represent the latest knowledge about various branches of the evolutionary tree of life.

46. W. H. Wagner Jr., "Origin and Philosophy of the Groundplan-divergence Method of Cladistics," *Systematic Botany* 5 (1980): 174–75.

47. P. E. Griffiths, "Cladistic Classification and Functional Explanation," *Philosophy of Science* 61 (1994), 216. Similarly, reports on recent fossil finds in China's Liaoning province have brought forth recollections of the predictions made several decades ago that dinosaur fossils would someday reveal a species with featherlike skin covering. "Chinese Site Yields Fossil Trove," *Boston Globe,* 5 May 1997.

48. Warren H. Wagner, "Applications of the Concepts of Groundplan-Divergence," in *Cladistics: Perspectives on the Reconstruction of Evolutionary Theory,* ed. Thomas Duncan and Tod F. Stuessy (New York: Columbia University Press, 1984), 101.

49. William B. Carpenter, "The Argument from Design in the Organic World" (1884), in his *Nature and Man* (London, 1888), 450; Bowler, *Evolution,* 225.

50. *Encyclopaedia Britannica,* 11th ed. (1910–11), s.v., "Paleontology" Romanes, *Darwin and After* vol. 1, chap. 5, "Paleontology"; T. H. Huxley, "On the Animals Which are Most Nearly Intermediate Between Birds and Reptiles," *Geological Magazine* (1868), reprinted in *Scientific Memoirs of Thomas Henry Huxley,* 308, 313; Bowler, *Evolution,* 202–4.

51. [F. W. Farrar], "Professor Max Müller's Lectures on Language: Second Series," *Reader* 4 (20 Aug. 1864): 221.

52. Oscar Schmidt, *The Doctrine of Descent and Darwinism* (New York, 1875), 303–10, quotation at 304. Schmidt's book formed part of the highly regarded International Scientific Series. Edited by E. L. Youmans, the series already included such well-known works as Walter Bagehot's *Physics and Politics* (1872), Herbert Spencer's *Study of Sociology* (1874), and John W. Draper's *History of the Conflict between Religion and Science* (1875).

53. Schmidt, *Doctrine of Descent,* 304.

54. Ibid., 249. J. Schmidt countered Schleicher's *Stammbaum* theory with a wave theory, which saw linguistic influence as radiating from a population center, effecting changes among neighboring language groups. This theory tried to account for the finding that the two great Indo-European subdivisions, Asian and European, were not as thoroughly cut off from each other as had once been supposed. Instead of a model of branching, a once and for all event, Johannes Schmidt proposed that the Indo-European dialects were related only according to geographical proximity, each an intermediate gradation between its neighbors on either side. Schmidt described changes in these languages by migration, like the spreading of waves from the point at which water is disturbed. N. E. Collinge, "Schmidt, Johannes," *The Encyclopedia of Language and Linguistics,* ed. R. E. Asher, 8 vols. (New York: Pergamon Press, 1994), 7:3681–82. For a defense of Schleicher's theory, see W. D. Whitney, "On Johannes Schmidt's New Theory of the Relationship of Indo-European Languages," *Journal of the American Oriental Society* 10 (1873): lxxvii–lxxviii.

55. Schmidt, *Doctrine of Descent,* 249–50.

56. The idea of a continuing emphasis on linguistic reconstruction may appear puzzling in light of the reputed shift from historical to structuralist linguistics in the decades after 1900. Yet the activities of, for example, the Linguistic Society of America (founded 1924)

show that, in practice, the early twentieth century has seen a significant continuation of historical linguistics / philology. Dell Hymes and John Fought, *American Structuralism* (New York: Mouton, 1981), 50–56.

57. W. Robertson Smith, "The Bible," *Encyclopaedia Britannica*, 9th ed. (1881); and other writings during the 1880s.

58. Ronald E. Clements, *A Century of Old Testament Study* (Guildford, England: Lutterworth Press, 1976), 1–7. The Q hypothesis was based on the widespread assumption of Markan priority: material in Matthew not found in Luke appeared to come largely from Mark. *The Anchor Bible Dictionary*, ed. David Noel Freedman (New York: Doubleday, 1990), s.v. "Synoptic Problem," and "Source Criticism."

59. On the science of "stemmatics": Manuscript copies of an ancient text were always many times removed from the lost original, the "autograph," yet scholars could arrange these into a *stemma codicum* or family tree of texts, and so follow back the threads of textual transmission with the goal of restoring the text as closely as possible to its original form. L. D. Reynolds and N. G. Wilson, *Scribes and Scholars*, 3d ed. (Oxford: Clarendon Press, 1991), 207–11.

60. *Encyclopaedia Britannica*, 11th ed. (1910–11), s.v.v. "Cuneiform," and "Hittite." For context, see chap. 1 of Bruce Kuklick, *Puritans in Babylon: American Archaeology and the Construction of the Ancient Near East* (Princeton: Princeton University Press, 1996).

61. Trautmann, "Revolution," 388–89 (see chap. 3, n. 11). Branching ethnology was much closer to Darwinism in spirit than was the new "evolutionist" anthropology, for it focused as much on descent as on development, and they shared the same kind of comparative method as distinct from the version used by E. B. Tylor and other sociocultural developmentalists.

62. Charles Neaves, *A Glance at some of the Principles of Comparative Philology* (Edinburgh, 1870), 7–8, 28–29.

63. Trautmann, *Morgan*, 182–91 (see Prologue, n. 4). Henry Maine reaffirmed the comparative-genealogical thesis on several occasions. Maine, *The Effects of Observation of India on Modern European Thought* (London, 1875); [H. S. Maine,] "The Patriarchal Theory," *Quarterly Review* 162 (1886): 197–98.

64. Trautmann, *Morgan*.

65. See George W. Cox, *The Mythology of the Aryan Nations* (London, 1870); Cox, *An Introduction to the Science of Comparative Mythology and Folklore* (London, 1881).

66. William Jones, quoted in frontispiece, Jaan Puhvel, *Comparative Mythology* (Baltimore: Johns Hopkins University Press, 1987). According to Puhvel (3–4, 13–14), Max Müller's genealogical view of the spread of myths is still in use today, even if his particular doctrines have long since been abandoned.

67. George E. Roberts, review of *Curiosities of Indo-European Tradition and Folk-Lore*, by Walter K. Kelly, *Reader* 2 (31 Oct. 1863): 499–500.

68. V. Gordon Childe, *The Aryans: A Study of Indo-European Origins* (New York: Knopf, 1926), 78–79.

69. *Encyclopedia of Language and Linguistics* s.v. "Paleontology, Linguistic"; Whitney, *Language and the Study of Language*, 205–8. The full title of Schrader's book was *Prehistoric Antiquities of the Aryan Peoples: A Manual of Comparative Philology and the Earliest Culture*.

70. Morris Swadesh introduced "Glottochronology," an attempt to fix dates on conjectured linguistic events. Inspired by carbon-14 dating of fossils, Swadesh devised a way to

measure the rate of lexical change or "decay" within a language: the idea was that the time depth separating two genetically related languages could be estimated by the degree to which they shared commonly inherited words. Swadesh, "Lexico-Statistic Dating of Prehistoric Ethnic Contacts," *Proceedings of the American Philosophical Society* 96, 4 (1952): 463. In 1969, Benveniste introduced "lexical semantics," the construction not of a mere protolexicon but of a protothesaurus, using clusters of terms to flesh out entire cultural categories. Emile Benveniste, preface to *Indo-European Language and Society,* trans. Elizabeth Palmer (Coral Gables, Fla.: University of Miami Press, 1973), 9–14.

71. Benjamin Ide Wheeler, "Indo-European Languages," *Johnson's Universal Cyclopedia,* 4:569 (see above, n. 27); Sievers, "Philology, ii, 18:786. A useful summary of the *Stammbaum* critique and the alternative theories appeared in Willem L. Graff, *Language and Languages: An Introduction to Linguistics* (New York: D. Appleton, 1932), 358–65.

72. As the American linguist Leonard Bloomfield later observed, "The earlier students of Indo-European did not realize that the family-tree diagram was merely a statement of their method: they accepted the uniform parent languages and their sudden and clear-cut splitting, as historical realities." Bloomfield, *Language* (New York: Henry Holt and Co., 1933), 311. Yet if the nineteenth-century comparativists were so deceived, their method was no less successful—which is the point stressed here.

73. A. H. Sayce, *Introduction to the Science of Language,* 2 vols. (London, 1880), 51; Sievers, Philology ii; T. G. Tucker, *Introduction to the Natural History of Language* (London, 1908), 67, 212; Renfrew, *Archaeology and Language,* 47, 67–69; W. S. Allan, "Comparative Reconstruction," *Encyclopedia of Language and Linguistics,* 2:636–43, esp. 642. Contemporary linguists such as Henry Hoenigswald and Larry Trask describe the various failings of the genealogical tree concept yet affirm that it has proven a useful enough guide in heuristic practice, by easily representing the findings of historical linguistics in etymology and language classification. Hoenigswald, "Language Family Trees," 257–59 (see chap. 2, n. 12); R. L. Trask, *Historical Linguistics* (New York: Arnold, 1996), 183–87. One change has occurred, however: treelike diagrams have sometimes been replaced by rectilinear charts. Though themselves highly schematic, these are at least less suggestive of complete severance of communication between two or more linguistic communities once past the point of mutual departure.

74. Haeckel's *History of Creation* devoted several chapters to forerunners of Darwin's transmutation theory. One of these, the early-nineteenth-century German geologist Leopold Buch, made a study of the Canary Islands which anticipated Darwin's stress on geographic isolation. Buch saw that isolated biological populations eventually became distinct species, and he compared this phenomenon to the way the political or topographical isolation of a human population causes its language to become distinct from its predecessors. Haeckel praised Buch for pointing out this "exceedingly instructive comparison . . . which is of the greatest use to the comparative-biological sciences." Haeckel, *History of Creation,* 107–8.

Recently, some researchers have sought a partnership between historical linguistics and the ethnogenetic history of humankind, yet this is but a small and highly controversial field. For this "coevolution" thesis, see L. L. Cavalli-Sforza, "Genetic and Linguistic Evolution," *Science* 244 (9 June 1989): 1128–29; David Penny, Elizabeth E. Watson, and Michael A. Steel, "Trees from Languages and Genes Are Very Similar," *Systematic Biology* 42 (1993): 382–84. Critiques of this expanded cladistic project abound: representative works include John H. Moore, "Putting Anthropology Back Together Again: The Ethnogenetic Critique of Cladistic Theory," *American Anthropologist* 96 (1994): 925–48; Trask, *Historical Linguistics,* 387–88, 401–2.

75. E. O. Wiley, "Karl R. Popper, Systematics, and Classification: A Reply to Walter Bock and Other Evolutionary Taxonomists," *Systematic Zoology* 24 (1975): 233–43; G. F. Engelmann and E. O. Wiley, "The Place of Ancestor-Descendant Relationships in Phylogeny Reconstruction," ibid. 26 (1977): 1–11. Platnick and Cameron argue that cladistic method generally "conforms to the dictates of Popperian 'hypothetical deductive' science," and suggest that this applies as much in historical linguistics as in biology. Norman Platnick and H. Don Cameron, "Cladistic Methods in Textual, Linguistic, and Phylogenetic Analysis," *Systematic Zoology* 26 (1977): 380–85.

76. *Encyclopedia of Language and Linguistics,* s.v. "Saussure, Ferdinand," "Hittite."

77. Ibid., s.v. "Linear B," "Indo-European Languages." This and the previous example are suggested in Wiener, "Of Phonetics and Genetics," 221.

78. *Encyclopedia of Language and Linguistics,* s.v. "Comparative Reconstruction."

ESSAY ON SOURCES

The notes to each chapter of this study cite the primary sources used as well as any secondary works that pertain to that chapter only. This bibliographic essay covers what remains: it describes the secondary historical literature I have used for background to the work as a whole. Works on the history of science appear first, then those on the history of linguistics. A handful of studies that self-consciously bring these fields together come at the end.

Most of the scientific figures in my story have been the subjects of biographies. On Darwin, the best and most recent works are Janet Browne, *Charles Darwin: Voyaging*, vol. 1 (New York: Knopf, 1995); and Adrian Desmond and James Moore, *Darwin* (London: Penguin, 1992). Browne's study, which is still in progress, is the more detailed of the two; Desmond and Moore's offers the benefit of being complete in one volume. No work interprets the whole of Charles Lyell's life and thought. Leonard G. Wilson, *Charles Lyell, the Years to 1841: The Revolution in Geology* (New Haven: Yale University Press, 1972), groans under the weight of scientific detail. Deservedly standard biographies of other figures include: A. Hunter Dupree, *Asa Gray* (Cambridge: Belknap Press, Harvard University Press, 1959); Edward Lurie, *Louis Agassiz: A Life in Science* (Chicago: University of Chicago Press, 1960); and Mario A. di Gregorio, *T. H. Huxley's Place in Natural Science* (New Haven: Yale University Press, 1984). Each of these provides a clear integration of its subject's scientific thought and lived experience.

Darwin scholarship has flourished in recent decades, helped along by the monumental labors of the "Darwin industry." The latter has made a host of primary sources available in published form: Darwin's corpus of writings, including his private manuscripts; his notebooks and book marginalia; and his complete correspondence, still in progress. These meticulously edited works reveal Darwin's personal life and his science in rich detail, offering researchers material for continued probing and reevaluation of his thought. For an introduction to the range of recent approaches and interpretations thus yielded, see David Kohn, ed., *The Darwinian Heritage* (Princeton: Princeton University Press, 1985). Other specialized studies on Darwin and Darwinism may be found in history of science journals, particularly the *Journal of the History of Biology*. Still admirable as a synthetic overview is Michael Ruse, *The Darwinian Revolution* (Chicago: University of Chicago Press, 1979). Also useful as an introduction to Darwin's science, although less historically oriented than Ruse's work, is Ernst Mayr, *One Long Argument: Charles Darwin and the Genesis of Modern Evolutionary Thought* (Cambridge: Harvard University Press, 1991). Peter Bowler, *Evolution: The History of an Idea*, rev. ed. (Berkeley: University of California Press, 1989), is the best single work setting Darwinism in its long-term intellectual context.

Background on a number of the issues treated in my own study appears in Howard E. Gruber, *Darwin on Man* (New York: E. P. Dutton, 1974), a work covering more of Darwin's thought than its title suggests; and Gruber, "Darwin's 'Tree of Nature' and Other Images of Wide Scope," in *On Aesthetics in Science*, ed. Judith Wechsler (Cambridge, Mass.: MIT Press, 1978), 121–40.

Historians of science have considered various aspects of Charles Lyell's thought, if not his experience. Martin Rudwick provides useful commentary on Lyell's magnum opus in his introduction to Lyell, *Principles of Geology*, 1st ed., ed. Rudwick, 3 vols. (1830–33; facsimile, Chicago: University of Chicago Press, 1990); also useful is Rudwick, *The Meaning of Fossils: Episodes in the History of Paleontology* (New York: Science History Publications, 1976). Michael Bartholomew, "Lyell and Evolution: An Account of Lyell's Response to the Prospect of an Evolutionary Ancestry for Man," *British Journal for the History of Science* 6 (June 1973): 261–303, gives a nuanced interpretation of Lyell's thinking, not only about human evolution but about Darwinism in general. On the interesting topic of Lyell's rhetoric as a constituent factor in his geological views, see Stephen J. Gould, *Time's Arrow, Time's Cycle: Myth and Metaphor in the Discovery of Geological Time* (Cambridge: Harvard University Press, 1987).

Nineteenth-century idealist morphology is an important topic for any discussion of the Darwinian revolution: Philip F. Rehbock, *The Philosophical Naturalists: Themes in Early Nineteenth-Century British Biology* (Madison: University of Wisconsin Press, 1983), treats the British appropriation of German *Naturwissenschaft*. Similarly, studies of Victorian natural science have rightly emphasized the era's theological apologetics. Ruse's *Darwinian Revolution* is useful on this subject, as are a number of shorter studies: Peter J. Bowler, for instance, analyzes the leading schools of natural theology in "Darwinism and the Argument from Design: Suggestions for a Reevaluation," *Journal of the History of Biology* 10 (Spring 1977): 29–43. The moral conundrums faced by natural theology are treated in Richard Yeo, "William Whewell, Natural Theology and the Philosophy of Science in Mid-Nineteenth Century Britain," *Annals of Science* 36 (1979): 493–516; and in several essays in Robert M. Young, *Darwin's Metaphor: Nature's Place in Victorian Culture* (Cambridge: Cambridge University Press, 1985).

An impressive literature on the history of linguistics has sprung up in recent decades, although biographically oriented studies are rarer here than in the history of science. The one relevant biography is Nirad C. Chaudhuri, *Scholar Extraordinary: The Life of Professor the Rt. Hon. Friedrich Max Müller* (New York: Oxford University Press, 1974), a work filled with detail yet reflecting, unfortunately, the spirit of its title. The best brief overview of Max Müller's life and thought appears in George W. Stocking Jr., *Victorian Anthropology* (New York: Free Press, 1987), 56–62. August Schleicher is treated in several of the essays in Hans Aarsleff, *From Locke to Saussure: Essays in the Study of Language and Intellectual History* (Minneapolis: University of Minnesota Press, 1982), and in a number of works by E. F. Konrad Koerner: see especially Koerner, *Practicing Linguistic Historiography: Selected Essays* (Philadelphia: John Benjamins, 1989); and his *Professing Linguistic Historiography* (Philadelphia: John Benjamins, 1995). Schleicher's linguistic *Stammbaum* logic and its historical background are discussed in Henry M. Hoenigswald, "Schleicher's Tree and Its Trunk," in *Ut Videam: Contributions to the Understanding of Linguistics*, ed. Werner Abraham (Lisse, Netherlands: Peter de Ridder, 1975). Material on Hensleigh Wedgwood, chiefly relating to his work on the *New English Dictionary*, can be found in Hans Aarsleff, *The Study of Language in England, 1780–1860* (Princeton: Princeton University Press, 1967). Little has been written about F. W. Farrar, who was not a major figure in the development of the linguistic field. Besides the article in the *Dictionary of National Biography,* nearly the only source is Reginald Farrar, ed., *The Life of Frederic William Farrar* (New York, 1904).

The standard surveys of nineteenth-century European linguistics are Otto Jespersen, *Language: Its Nature, Development and Origin* (1922; reprint, London: Allen & Unwin, 1959); and Holgar Pedersen, *The Discovery of Language: Linguistic Science in the Nineteenth Century,* trans. John W. Spargo (1924; reprint, Bloomington: Indiana University Press, 1931). Although these works are useful for the wealth of information they provide, better for interpretation are more recent surveys: Robert H. Robins, *A Short History of Western Linguistics* (Bloomington: Indiana University Press, 1967); and Roy Harris and Talbot J. Taylor, *Landmarks in Linguistic Thought* (New York: Routledge, 1989). The latter work stresses episodes in the development of linguistic theory and is arranged by individual thinker. Hans Aarsleff has excelled in revising the traditional historiography of eighteenth- and nineteenth-century linguistics, one of his targets being the emphasis found in the Jespersen and Pedersen surveys. This older school stressed Germany to the neglect of other European countries and overemphasized the idea of a nineteenth-century revolution in comparative-historical method. On this subject see especially the introduction to Aarsleff's *From Locke to Saussure.* Two books by Thomas R. Trautmann further this revision of what we might call the modern-science historiography of nineteenth-century linguistic study. Trautmann, *Lewis Henry Morgan and the Invention of Kinship* (Berkeley: University of California Press, 1987); and Trautmann, *Aryans and British India* (Berkeley: University of California Press, 1997), address the wider intellectual and ideological context of comparative philology, especially the field's ethnological imperative. In particular, Trautmann details the way comparative philology and ethnology perpetuated the Mosaic genealogy of nations and, at least at first, the Bible's short ethnological time frame.

The major sources on Victorian ethnology are: Nancy L. Stepan, *The Idea of Race in Science: Great Britain, 1800–1960* (Hamden, Conn.: Archon Books, 1982); and Stocking, *Victorian Anthropology.* The standard intellectual history of the nineteenth-century American polygenist school is William Stanton, *The Leopard's Spots: Scientific Attitudes toward Race in America, 1815–1859* (Chicago: University of Chicago Press, 1960). For an overview of Indo-Europeanist ethnology, see J. P. Mallory, "A History of the Indo-European Problem," *Journal of Indo-European Studies* 1:1 (1973): 21–65. Maurice Olender provides intellectual-historical richness concerning the ideologies informing nineteenth-century linguistic and ethnological scholarship in *The Languages of Paradise: Race, Religion, and Philology in the Nineteenth Century,* trans. Arthur Goldhammer (Cambridge: Harvard University Press, 1992). On the eighteenth-century roots of the comparative method and the genealogical model in linguistics, see the essays by Henry M. Hoenigswald, Konrad Koerner, and R. H. Robins, in *Leibniz, Humboldt, and the Origins of Comparativism,* ed. Tullio de Mauro and Lia Formigari (Philadelphia: John Benjamins, 1990). A learned history of an important aspect of European linguistic thought may be found in James H. Stam, *Inquiries into the Origin of Language: The Fate of a Question* (New York: Harper & Row, 1976).

Historians primarily of language study, not of science, have begun recently to explore the relationship between linguistic and biological thought. The best single source on this topic is Henry Hoenigswald and Linda F. Wiener, eds., *Biological Metaphor and Cladistic Classification* (Philadelphia: University of Pennsylvania Press, 1987), especially the historical essays in the volume's first half. Additional observations on the idea of a parallel between linguistics and geology, particularly as conceived in the nineteenth century, appear in three works: Rulon Wells, "Uniformitarianism in Linguistics," in *Dictionary of the History of Ideas*, 4 vols., ed. Philip P. Wiener (New York: Charles Scribners, 1973), 4:423–31; T. Craig Christy, *Uniformitarianism in Linguistics* (Philadelphia: John Benjamins, 1983); and Bernd Naumann, Frans Plank, and Gottfried Hofbauer, eds., *Language and Earth: Elective Affinities between the Emerging Sciences of Linguistics and Geology* (Philadelphia: John Benjamins, 1992).

Much remains to be written about the genealogical vision in linguistic scholarship prior to the Darwinian era, including consideration of how this may have influenced the advent of genealogical thinking in biological systematics. On this topic, see W. Keith Percival's masterful essay "Biological Analogy in the Study of Language Before the Advent of Comparative Grammar," in *Biological Metaphor and Cladistic Classification*, ed. Hoenigswald and Wiener. See also T. M. S. Priestly, "Schleicher, Čelakovsky, and the Family-Tree Diagram," *Historiographia Linguistica* 2 (1975): 299–333.

INDEX

Page numbers in italics indicate illustrations.

Aarsleff, Hans, 150n. 8

Academy, 89

Aesthetics in science, 1, 147, 148

Agassiz, Louis, 4, 18, 36, 44, 47, 57, 108, 154n. 17, 161n. 18, 163n. 50, 169n. 112, 175n. 89; career, 40; language-species analogy, 50, 51; on linguistic genealogy, 41–42; marginalia in Lyell's *Antiquity of Man,* 163n. 51

Albright, William F., 131

Allan, W. S., 145

Allen, David Elliston, 152n. 10

American Academy of Arts and Sciences, 47, 50–51

American Museum of Natural History, 126

Archaeology, 26, 27

Argyll, Duke of, 98, 103, 105, 174n. 73

Aryan. *See* Indo-European family

Athenaeum, 68, 88, 89, 96, 134

Atlantic Monthly, 47, 48, 67, 97

Baden Powell, Baden, 45, 62, 162n. 39

Barker, T. Childe, 141

Bartholomew, Michael, 26

Beagle, 12, 15

Beer, Gillian, 150n. 10; 156nn. 29, 34; 175n. 85

Bible, 7; Mosaic ethnology, 30; New Testament gospels, 140; Pentateuch, 140; place names, 131; Tower of Babel, 7

Bikkers, Alex V. W., 172n. 45

Birge, E. A., 177n. 27

Bishop analogy, 21–24, 28, 52, 53, 164n. 56

Bloomfield, Leonard, 182n. 72

Bopp, Franz, 10, 18, 32, 83, 157n. 44

Bowen, Francis, 66n. 89

Bowler, Peter, 5, 6, 150nn. 9, 12, 153n. 11, 154n. 11, 158n. 57, 173n. 53, 174n. 72, 176n. 10, 178n. 31

Bridgwater Treatises, 45

British Association for the Advancement of Science, 42, 43, 89, 98

Brixham cave, 39

Bronn, Heinrich, 73

Browne, Janet, 156n. 36

Buch, Leopold, 182n. 74

Buffon, Comte de, 9

Bunsen, C. C. J., 39, 43

Burnett, James (Lord Monboddo), 16, 153n. 4

Burrow, J. W., 150n. 8

Cambridge Philological Society, 87

Cambridge University, 11, 86

Cameron, H. Don, 183n. 75

Cannon, Susan Faye (Walter F.), 3

Carpenter, William B., 138, 160n. 9, 169n. 111

Cavalli-Sforza, L. L., 182n. 74

Chambers, Robert, 6, 32, 160n. 9. See also *Vestiges of the Natural History of Creation*

Childe, V. Gordon, 142

Christian Examiner, 41

Cladistic classification, 136–37, 179n. 45

Classification of languages: genealogical, 28–30, 102; grammatical, 8, 42, 92, 102

Common descent: biological, 5, 6, 18, 20, 21, 49, 102, 112, 113, 114, 115, 121, 123–28, 136–40; in Darwin's *Origin:* 28, 29, 75; linguistic, 17, 87, 92, 102, 128–34. *See also* Indo-European family, linguistic

Comparative anatomy, 9

Comparative method: in human sciences, 141–42; in linguistics, 2, 9–10, 109, 136

Comte, Auguste, 87

Conjectural history, 33, 136, 159n. 64

Conser, Walter, 161n. 19

Cope, Edward Drinker, 125

Cornhill Magazine, 49

Coulanges, Fustel de, 141

Cox, George Wilson, 141, 171n. 29

Craw, Robin, 178n. 29

Cuvier, Georges, 9, 13, 97, 151n. 5

Darwin, Charles: Agassiz, 51; and conjectural history, 33; continuity with Lyell, 25, 27–28, 104; ethnological reading, 31, 37–38; on Farrar, 100; function of analogies, 2, 3; on genealogical classification, 20, 28; and Gray, 47, 48, 53–56, 64–68, 99; and Herschel, 12; on human evolution, 56, 65, 96–97; and Lyell, 43, 64–68, 100; on Max Müller, 54, 100; notebooks, 15–17, 19, 25; on the origin of language, 16–17, 52, 100; on Owen, 97–98; on racial monogenesis, 37–38; on ridicule, 22; on

Library of Congress Cataloging-in-Publication Data

Alter Stephen G.

    Darwinism and the linguistic image : language, race, and natural theology
in the nineteenth century / Stephen G. Alter.

      p.   cm. — (New studies in American intellectual and cultural history)
    Includes bibliographical references and index.

    ISBN 0-8018-5882-8 (alk. paper)

    1. Linguistics—History—19th century.   2. Comparative linguistics.
3. Darwin, Charles, 1809–1882—Influence.   4. Evolution (Biology)
5. Biolinguistics.   I. Title.   II. Series.

P75.A47   1998

410'.9'09034—dc21

                                                           98-14270
                                                               CIP